CULTURE AND COSMOS
http://www.CultureAndCosmos.org

Culture and Cosmos is published twice a yea
autumn/winter, in association with the Sophi:
Culture, University of Wales Trinity Saint D&
Contributions and editorial correspondence sl
Editors@cultureandcosmos.org

Editor: Dr Nicholas Campion, the Editor of *Culture and Cosmos*, University of Wales Trinity Saint David, Lampeter, Ceredigion, Wales, SA48 7ED, UK.
E Mail **n.campion@uwtsd.ac.uk**
Deputy Editor: Dr Jennifer Zahrt
Editorial Board: Dr Silke Ackermann, Professor Anthony F. Aveni, Dr David Brown, Professor Charles Burnett, Dr Hilary M. Carey, Dr John Carlson, Dr Patrick Curry Professor Robert Ellwood, Dr Germana Ernst, Dr Ann Geneva, Professor Joscelyn Godwin, Dr Dorian Greenbaum, Dr Jacques Halbronn, Dr Robert Hand, Dr Jarita Holbrook, Professor Michael Hunter, Professor Ronald Hutton, Dr Peter Kingsley, Dr Edwin C. Krupp, Dr J. Lee Lehman, Dr Lester Ness, Professor P. M. Rattansi, Professor James Santucci, Robert Schmidt, Dr Fabio Silva, Dr Lorenzo Smerillo, Professor Richard Tarnas, Dr Graeme Tobyn, Dr David Ulansey, Robin Waterfield, Dr Charles Webster, Dr Graziella Federici Vescovini, Dr Angela Voss, Dr Paola Zambelli, Robert Zoller.
Copy Editor Kathleen White
Web: Dr Frances Clynes

Contributors Guidelines: Please see http://www.cultureandcosmos.org/submissions.html

Front cover: The LDS Temple in San Diego. See Shon Hopkin, 'The Joining of Heaven and Earth in Mormon Sacred Texts and Temples'.

Published by Culture and Cosmos, Dr Nicholas Campion, Faculty of Humanities and the Performing Arts, University of Wales Trinity Saint David, Lampeter, Ceredigion, Wales, SA48 7ED, UK.

© **Culture and Cosmos 2017**
Printed by Lightning Source.

The Sophia Centre
http://www.uwtsd.ac.uk/sophia

The Centre for the Study of Cosmology in Culture is an academic centre within the Faculty of Humanities and the Performing Arts at the University of Wales Trinity Saint David.

The Centre's academic goals are

- 'to pursue research, scholarship and teaching in the relationship between astrological, astronomical and cosmological beliefs and theories, and society, politics, religion and the arts, past and present' and
- 'to undertake the academic and critical examination of astrology and its practice'.

The Centre's wider goal is stated in its title – to 'study cosmology in culture'. In a traditional sense, a cosmology is a world view, an understanding of the cosmos which informs individual and social action and ideology. The Centre promotes research in the subject area, holds seminars and conferences, publishes scholarly material, is associated with Sophia Centre Press and supervises PhD students.

The Centre's teaching is focused on the MA Cultural Astronomy and Astrology. For further information see
http://www.uwtsd.ac.uk/ma-cultural-astronomy-astrology

THE MARRIAGE OF HEAVEN AND EARTH

A special issue of *Culture and Cosmos*
Vol. 20 nos. 1 and 2
Spring/Summer and Autumn/Winter 2016

Published by Culture and Cosmos
& Sophia Centre Press
England

www.cultureandcosmos.org

In association with the
Sophia Centre for the Study of Cosmology in Culture,

University of Wales Trinity Saint David,
Faculty of Humanities and the Performing Arts
Lampeter, Ceredigion, Wales, SA48 7ED, UK

British Library Cataloguing in Publication Data
A catalogue card for this book is available from the British Library

ISSN 1368-6534

Printed in Great Britain by Lightning Source

CONTENTS

Alexander Cummins
The Worldly Faces of the Heavens:
Nature and Seventeenth-Century English Astrological Images

CULTURE AND COSMOS

www.CultureAndCosmos.org

Editor Nicholas Campion
Vol. 20 No. 1 and 2 Autumn/Winter and Spring/Summer 2016
ISSN 1368-6534

Published in Association with
The Sophia Centre for the Study of Culture in Cosmology,
Faculty of Humanities and the Performing Arts
University of Wales Trinity Saint David
http://www.uwtsd.ac.uk/sophia/

Editorial

This issue of *Culture and Cosmos* includes edited papers based on presentations at the 2014 Sophia Centre Conference titled 'The Marriage of Heaven and Earth': the notion of marriage as the creative and productive union of archetypal forces, powers or entities is borrowed from the alchemical tradition. As is in previous years the conference explored a range of topics covered by the umbrella term 'cultural astronomy and astrology', as represented through the research and teaching of the Sophia Centre University of Wales Trinity Saint David, including the MA in Cultural Astronomy and Astrology.

The choice of words – Heaven and Earth – was intended to evoke religious imagery. The text in the call for papers made this clear: 'All human cultures have both identified the sacred in the landscape, and created structures which embody the sacred. In many cases these sacred spaces are related to the stars, planets and sky. This academic conference will consider the construction, creation and representation of the sky in sacred space'.

I have divided the papers which were submitted and passed the review process into two themes. The first, 'Land and Architecture', crosses time periods and cultures. It opens with Juan Antonio Belmonte's study of 'Cosmic Landscapes in Ancient Egypt', and then moves to four papers on the Americas. We begin in the south with Kim Malville's 'Passages between Worlds: Heaven, Earth, and the Underworld in the Andean Cosmos' and then move north to Mexico with two papers: Hal Green's study of 'The Zenith Sun as an Organizing Principle of the Constructed

Sacred Space and Calendrics of Central Mexico' is followed by Stanisław Iwaniszewski's paper on 'Communicating with the Ancestors in the Spiritual Landscape at Yaxchilán, Chiapas, Mexico'. We continue our journey to the north and conclude this section with Shon Hopkin's examination of 'The Joining of Heaven and Earth in Mormon Sacred Texts and Temples'.

The second group of papers are gathered under the title 'Text and Image'. As in our section on 'Land and Architecture', we open with Egypt, and with Joanna Popielska-Grzybowska's 'Some Remarks on the Sky in the Ancient Egyptian Pyramid Texts'. The next four papers all deal with medieval and Renaissance Europe. Scott Hendrix has written a wide-ranging examination of the theological implications and controversies of astrology in the Catholic world, Gerardina Antelmi takes us to literature with her study of 'Poetry Creation as Space of Union between Natural and Supernatural: A Reading of *The House of Fame*', Edina Eszenyi explores 'Shaping the Image of Lucifer in the Cinquecento Veneto' and Alexander Cummins concludes with his paper on 'The Worldly Faces of the Heavens: Nature and Seventeenth-Century English Astrological Images'.

I would like to thank all the contributors for presenting papers at the conference, submitting drafts for publication and then coping patiently with reviewers' and editorial requests. Thanks are also due to the University of Wales Trinity Saint David for its ongoing support, including our Dean in the Faculty of Humanities and the Performing Arts, Dr Jeremy Smith, and Assistant Dean, Dr Kyle Erickson. Also thanks to our copy-editor Kate White, and Deputy Editor Dr Jenn Zahrt for seeing this delayed issue through to publication.

Dr Nicholas Campion,
Sophia Centre for the Study of Culture in Cosmology,
Faculty of Humanities and the Performing Arts,
University of Wales Trinity Saint David.

Cosmic Landscapes in Ancient Egypt:
A Diachronic Perspective

Juan Antonio Belmonte

Abstract: The walls of the temples of the goddess Hathor at Denderah and Horus at Edfu have offered many representations of the *stretching of the cord* ceremony where the northern, *imperishable*, ancient Egyptian constellation of Meskhetyu is advocated as the target that the Pharaoh was looking at when fixing the orientation of the temples. However, it is often asserted that, most commonly, temples built along the Nile were oriented on an axis according to local topography as determined by the river. Our decade of work in Egypt has proven that both possibilities are not only true but complementary. From the onset of their civilization, ancient Egyptians selected certain sectors of their geography, blessed by peculiar orographic and celestial conditions, with the aim of creating cosmic landscapes. This tradition can be traced from dawn to dusk of their civilization in the proto-Dynastic and Roman periods, respectively. In this essay review, several of the most sacred places of the Black Land will be visited, following a diachronic perspective that will show how Abydos, Memphis, Thebes and Denderah, among others, were transformed into real cosmic landscapes, reflecting the Egyptian worldview of each epoch: from the astral eschatology of the Pyramid Texts to the highlights of the solar cult as represented by Karnak.

Introduction

In this review paper we are going to deal with precisely the relationship between astronomy and landscape, and the role that the observation of the sky could play, not only in the orientation of a particular monument but also in the choice of the location, if not in a combination of both. This phenomenology will be related to certain other aspects of the Pharaonic civilization as the calendar,[1] the mapping of the sky,[2] and the celestial aspects of the religion.[3]

---

[1] See J. A. Belmonte, 'Some Open Questions on the Egyptian Calendar: An Astronomer's View', *Trabajos de Egiptología (Papers on Ancient Egypt)* 2 (2003): pp. 7–56; and L. Depuydt, 'Calendars and Years in Ancient Egypt: The Soundness of Egyptian and West Asian Chronology in 1500–500 B.C. and the Consistency of the Egyptian 365-day Wandering Year', in *Calendars and Years:*

Juan Antonio Belmonte, 'Cosmic Landscapes in Ancient Egypt: A Diachronic Perspective', *The Marriage of Heaven and Earth,* a special issue of *Culture and Cosmos*, Vol. 20, nos. 1 and 2, 2016, pp. 3–30.
www.CultureAndCosmos.org

The archaeology of the landscape – the discipline in which frame we are going to move around – has not been a subject that has enjoyed great prestige in the studies of Egyptology; only recently a timid approach to the discipline has begun to be developed, such as the study of ritual landscape in ancient Thebes or our own work throughout the country.[4] To date, most of the studies have been limited to a very prosaic relationship with the Nile, ignoring other elements of the landscape, such as the horizon, and of course, with very rare exceptions, the celestial phenomena occurring on it and the rest of the celestial vault.[5]

However, the author and his collaborators – mostly but not all within the context of the Egyptian-Spanish Mission on Archaeoastronomy of ancient Egypt – have dealt with a series of crucial analyses that have largely demonstrated the important role played by the heavens in the alignment of sacred buildings in ancient Egypt.[6] In this essay we are going to briefly analyse some of the most striking cases, in particular those in

---

Astronomy and Time in the Ancient Near East, edited by J. M. Steele, (Oxford: Oxbow Books, 2007), pp. 35–82, and references therein.

[2] J. Lull and J. A. Belmonte, 'The Constellations of Ancient Egypt', in *In Search of Cosmic Order, Selected Essays on Egyptian Archaeoastronomy*, edited by J. A. Belmonte and M. Shaltout, (Cairo: Supreme Council of Antiquities Press, 2009), pp. 155–94.

[3] R. Krauss, 'Astronomische Konzepte und Jenseitsvorstellungen in den Pyramidentexten', *Ägyptologische Abhandlung Band 59* (1997); H. Brugsch, *Thesaurus Inscriptionum Aegyptiacarum*, Vol. I: Astronomische und astrologische Inschriften altaegyptischer Denkmäler (Leipzig: J. C. Hinrichs, 1883).

[4] M. Ullmann, 'Thebes: origins of a ritual landscape', in *Sacred Space and Sacred Function in Ancient Thebes*, edited by P. F. Dorman and B. M. Bryan, (Chicago: The Oriental Institute of the University of Chicago, 2007), pp. 3–26; J. A. Belmonte, *Pirámides, templos y estrellas: astronomía y arqueología en el Egipto antiguo* (Barcelona: Crítica, 2012), pp. 215–50.

[5] R. H. Wilkinson, *The Complete Temples of Ancient Egypt* (London: Thames & Hudson, 2000), pp. 36–37; L. Gabolde, 'La date de fondation du temple de Sésostris Ier et l'orientation e l'axe', in *Le Grand Château d'Amon de Sésostris Ier à Karnak* (París: Parution, 1998); and L. Gabolde, 'L'horizon d'Aton, exactement?', *Cahiers de l'Égypte Nilotique et Mediterranéenne* 2 (2009): pp. 145–57.

[6] J. A. Belmonte, M. Shaltout, and M. Fekri, 'Astronomy, Landscape, and Symbolism: A Study on the Orientations of Ancient Egyptian Temples', in *In Search of Cosmic Order*, Belmonte and Shaltout (eds.), pp. 213–84.

which astronomy combined with local topography on a large scale, to fight chaos (*isfet*) and help maintain cosmic order (*ma'at*).[7]

We are going to focus on some of the most famous archaeological sites in Egypt: the archaic royal cemetery at Abydos, the pyramid fields, notably the 4th Dynasty behemoths at Dahshur and Giza, the complex of Karnak in Luxor, and the temple complex of the goddess Hathor at Denderah.[8]

History begins at Abydos.

The author's research team has proposed that structure HK29A at Kom Al-Ahmar (Hierakonpolis or ancient Nekhen) was one of the first buildings erected in Egypt that was astronomically orientated, and it is of course the oldest that has been found.[9] Hierakonpolis presumably was the city where the cult of the Falcon God Horus was developed and it is certain that the first kings of Egypt had close ties with the town. A peculiar configuration of the Egyptian imperishable constellation Meskhetyu would have been used in Kom Al-Ahmar to find directions, perhaps because it was not circumpolar on site during the Predynastic period, since its lowest star Merak (βUMa) slightly disappeared below the local horizon at that time.

However, Meskhetyu became entirely circumpolar a few hundred kilometres to the north of Nekhen, at a site that would transform into one of the most important sacred sites of Egypt: Abydos (see Fig. 1). Umm el Qab, in the desert area of Abydos, is the site of a huge Pre- and Proto-Dynastic cemetery which also includes the tombs of the first Kings of a unified Egypt (0th and 1st Dynasties), at the founding moment of the Egyptian state, when certain metaphysical aspects relating to the figure of

---

[7] E. Hornung, *El uno and los múltiples: concepciones egipcias de la divinidad* (Madrid: Trotta, 1999).

[8] For a detailed archaeological description of these sites, see: S. Aufrère, J. C. Golvin and J. C. Goyon, *L'Égypte restituée*, Volume 1, *Sites et temples de l'Haute Égypte* (París: Editions Errance, 1991); S. Aufrère and J. C. Golvin, *L'Égypte restituée*, Volume 3. *Sites, temples et pyramides de Moyenne at basse Égypte* (París: Editions Errance, 1997); M. Lehner, The complete pyramids (London: Thames & Hudson, 1997); and Wilkinson, *The Complete Temples of Ancient Egypt*, pp. 36–37.

[9] R. Friedman, 'The Ceremonial Centre at Hierakonpolis Locality HK29A', in *Aspects of Early Egypt*, ed. J. Spencer, (London: British Museum Press, 1996), pp. 16–35; J. A. Belmonte, M. Shaltout, and M. Fekri, 'On the Orientation of Ancient Egyptian Temples: (4) Epilogue at Serabit el Khadem and Overview', *Journal for the History of Astronomy* 39 (2008): pp. 181–211.

the King were first developed, perhaps including the stellar eschatology of the Pyramid Texts.[10]

Fig. 1. A cosmic landscape at Abydos (ancient Egyptian *Abdju*), the site of the burial monuments of the first kings of a unified Egypt. A distinct orographic feature on the local landscape – (a) the royal tombs at Umm el Qab, (b) tomb of Den and the funerary precincts, and (c) Khasekhemuy's 'Shunet el-Zebib') – were aligned in a south-north axis (dotted-line) perhaps pointing to the lower culmination of Merak (βUMa) – the lowest for any star in Meskhetyu (d) – as seen from the Royal Cemetery, where the 'imperishable' asterism was circumpolar during the First Dynasty. This might have justified the sacredness of the place. The tombs and the precincts themselves were inter-cardinally orientated (dashed-line). See the text for further details. Diagram by the author.

---

[10] For the Pyramid Texts, a classical work is R. O. Faulkner, *The Ancient Egyptian Pyramid Texts* (Oxford: Oxford University Press, 1969). For a most recent recession see: J. P. Allen and P. Der Manuelian, *The Ancient Egyptian Pyramid Texts. Writings from the Ancient World* (Ann Arbor, MI: Society of Biblical Literature Press, 2005).

Abydos is located at a very striking site in Middle Egypt. The hieroglyph for mountain in the ancient writing was ⌣, *dju*. Actually, there are few natural mountains in the Nile valley. Most of the orographic elements are the bluffs of the desert ridges, crossed by seasonal water streams or *wadis*. This orographic accident probably is the origin of this sign, which is extremely ancient, as it appears already in the seals found in the pre-dynastic tomb U-j at Umm el Qab, where a 'King Elephant mountain' is mentioned.[11] Its connection with the afterlife, and in particular with Abydos, is clear. To the southern horizon of the necropolis a huge *wadi* opens into Umm el-Qab bay, functioning as a sort of 'mouth of the afterworld' (see Fig. 1). The sign appears explicitly in the name of Abydos, which was:

, or *Abdju*.

Giulio Magli has presented a detailed analysis of the site, arguing the existence of a direct connection between the topographical feature, the necropolis at Umm el Qab and the funerary precincts at the limit of the cultivable lands (see Fig. x.1).[12] All of them are aligned within a meridian axis that suggests a certain relationship to the northern skies, as we will show below. The Royal Tombs and the associated funerary complexes, including the largest surviving early structure at Abydos (the precinct of King Khasekhemuy at Shunet el-Zebib, see Fig. 1), are distinct members of the inter-cardinal family of orientations (developed in Upper Egypt to accomplish temples that were aligned correctly both to the sky and to the Nile) which was possibly a particular case of a general cardinal pattern related to the northern stars.[13] However, Magli probably fails in his interpretation of inter-cardinal directions in Upper Egypt, notably in proto-Dynastic Abydos but also in New Kingdom Thebes. He argues that the idea that the Egyptian temples were mainly orientated towards the Nile tends to be overplayed in Egyptological literature; our hypothesis − as

---

[11] G. Dreyer, 'Umm el-Qaab 1: Das prädynastische Königsgrab U-j und seine frühen Schriftzeugnisse', *DAI Archäologische Veröffentlichungen* 86 (1998).
[12] G. Magli, *Architecture, astronomy and sacred landscape in ancient Egypt* (Cambridge: Cambridge University Press, 2013).
[13] J. A. Belmonte, M.A. Molinero, and N. Miranda, 'Unveiling Seshat: New Insights into the Stretching of the Cord Ceremony', in *In Search of Cosmic Order*, Belmonte and Shaltout (eds.), pp. 193–210. See also, J. A. Belmonte, M. Shaltout and M. Fekri, 'Astronomy, Landscape and Symbolism', pp. 213–84.

defended by our research group – is also an effort to reconcile Nile orientation with the astronomical one, an explanation which, according to Magli, would be too far from 'what we know about the Egyptian architect's way of thinking'. He finds a far more natural astronomical interpretation, possibly inspired by his earlier work on the Inka: the Milky Way. However, there is a lack of ancient texts supporting his ideas.[14]

Middle and especially New Kingdom Thebes replicate Abydos, the place of the ancestor burial. In Abydos, northeast orientations are dominant (see Fig. 1), while southeast is the preferred orientation in Thebes. While the Galaxy could explain the orientation in Abydos, only a custom of placing temple façades perpendicular to the Nile valley can explain both orientations simultaneously. Inter-cardinal directions are present, with minor exceptions, only in Upper Egypt, where the corresponding alignments would be nearly perpendicular to the Nile, and they are absent elsewhere in Egypt. This would not happen if the reference were a celestial body. Similarly, the orientation family peaks are rather sharp in the related histograms, something difficult to explain with an orientation to a large and loose object covering a wide range of declinations (such as the Milky Way).[15]

The question now is why the northern skies could be relevant at Abydos. Two scenarios are possible. Firstly, it could be argued that Umm el Qab was selected as the site for the early dynastic Royal Cemetery because it was the first place, following the northward flow of the Nile from Nekhen, where Meskhetyu would have been a circumpolar asterism (see Fig. 1, i.e. it behaved as an imperishable sky entity), plainly justifying the sacred character of Abydos and suggesting that the site was deliberately selected. On the contrary, the alternative hypothesis would be that the relative importance of Meskhetyu as the quintessential imperishable constellation was due to the fact that it was circumpolar at the site of the royal necropolis. Both solutions may be possible and certainly indistinguishable from the astronomical and topographic point of view.

However, the first option seems the most likely because it would represent a closer relationship between astronomy and geography with the creation of a true cosmic landscape. In addition, Toby Wilkinson has

---

[14] Magli, *Architecture, Astronomy and Sacred Landscape in Ancient Egypt*, pp. 14–24.

[15] Belmonte, *Pirámides, templos y estrellas*, Figure 11; see also, J. A. Belmonte and A.C. González-García, 'The Pillars of the Earth and the Sky: Capital Cities, Astronomy and Landscape', *Journal of Skyscape Archaeology* 1 (2015): pp. 8–37.

suggested that the founder Kings of Egypt (0th and 1st Dynasties) originally were from Nekhen, but that they established their Royal Necropolis a few hundred miles further to the north, at Umm el Qab and Abydos, for unknown reasons.[16] Our suggestion could support the idea that the place of the royal necropolis (and by extension Abydos) was deliberately chosen for religious reasons related to the astral eschatology. This would convert Abydos in the primeval Egyptian Cosmic Landscape later to be imitated in other royal sites of Egypt under similar but not necessarily identical frameworks.

Kha(wy) Sneferu: a 'balance' between the Two Lands

The two pyramids built during the Old Kingdom by the 4th Dynasty Pharaoh Sneferu at Dahshur are usually considered as two consecutive projects, the second – that of the Red Pyramid – being generated by a presumed failure of the first, the Bent Pyramid.[17] However, Belmonte and Magli have shown that the archaeological proofs of such a scenario are far from certain and that, on the contrary, a series of architectural, topographical, epigraphic and astronomical hints point to a unitary project probably conceived from the very beginning in terms of the two pyramids and of their annexes.[18] The two pyramids together are thus shown to form a conceptual, sacred landscape associated with the Pharaoh's afterlife.

The leitmotif of the ideas defended by Belmonte and Magli was the independent research on the topic performed by the two authors who, starting from different approaches – symbolism and architectural design, with a common interest in a cultural astronomy approach – were able to reach to the identical conclusion that the pair of gigantic pyramids at Dahshur were finished, and even perhaps conceived, as a unique project where, apart from the strong symbolic aspect and the obligation of accomplishing king's necessities for the hereafter, Sneferu was able to show all his dominion of the land by the simple contemplation of his monuments.

From Sneferu onward, if not earlier, each royal pyramid complex will receive a name, which in several cases has been passed on to us in the

---

[16] T. A. H. Wilkinson, *Early Dynastic Egypt* (New York: Routledge, 1999).

[17] A. Fakhry, *The Monuments of Sneferu at Dahshur* (Cairo: General Organization for Govt. Print Offices, 1959). M. Lehner, *The Complete Pyramids* (London: Thames & Hudson, 1997), pp. 97–107.

[18] J. A. Belmonte and G. Magli, 'Astronomy, Architecture and Symbolism: The Global Project of Sneferu at Dahshur', *Journal for the History of Astronomy* 46 (2015): pp. 173–205.

reliefs of the tombs of the officials and priest in charge of the complexes.[19] The Red Pyramid was called ⟨hieroglyphs⟩. The name is made up of Sneferu's name as Egyptian Double King, inscribed in a cartouche – the hieroglyph ⟨sign⟩ – *kha* and the pyramid determinative – and should be read as Kha-Sneferu, which is usually translated as 'Sneferu shines' or 'Sneferu is bright'. However, the verbal stem *kha* describes both the rising (and hence their first brightness) of a celestial object and the king's assumption of royal regalia, notably of the crowns, as well as his ritual appearances in festival and processions.[20] Interestingly, the ancient name of the Bent Pyramid was the same as that of the Red Pyramid, ⟨hieroglyphs⟩, but with the addition of a sign denoting 'south' or also 'Upper Egypt'. This would be read as Kha-Sneferu-Resy, and has been standardly translated as 'Sneferu shines in the south'. However, considering the previous discussion, the Bent Pyramid could be translated as 'Sneferu wearing the crown of the south' or, directly, the 'Southern Crown of Sneferu'.

The similarity, or complementariness, of the names of the two pyramids is certainly another hint for a common project. Actually, if we seek further support, we might note that Sneferu's mortuary complex was referred to in a decree issued by the 6th Dynasty king Pepi I (ca. 2250 BCE) regulating administration of the pyramid town in Dahshur, found on an inscribed stone not far from the Red Pyramid, as ⟨hieroglyphs⟩, Kha(wy) Sneferu, which could be translated either as 'The two pyramids – Sneferu shines', or perhaps, even better as 'The two crowns of Sneferu'.[21] It is our contention that the hypothesis that the two great pyramids of Sneferu at Dahshur were conceived as a unitary project is indeed viable and most likely true. In ⟨hieroglyphs⟩, astronomy, religion, politics and architecture combined to produce a tangible reality charged with symbolism on a truly gigantic scale.

The idea – earlier expressed by Cintron on the evidence based on the Palermo Stone – is that the two pyramids of Sneferu at Dahshur – which could possibly and reasonably be identified as huge symbolic representations of the upper Egypt White Crown for the Bent Pyramid ⟨hieroglyphs⟩, and of the Lower Egypt Red Crown for the Red one

---

[19] S. Quirke, *Ra, el Dios del Sol* (Madrid: Anaya-Spain, 2003), p. 145.

[20] K. Goebs, *Crowns in Egyptian Funerary Literature: Royalty, Rebirth, and Destruction* (Oxford: Griffith Institute Monographs 2008), pp. 24–31.

[21] A. Moret, 'Chartes d'immunité dans l'Ancien Empire égyptien', *Journal Asiatique* 20 (1912): pp. 359–447.

(𓉐𓊹𓈖𓇯𓈋) 𓎛𓊃 – were at the same time the materialization in white and red stone, respectively, of celestial realities that were most relevant for the astralization, and hence the eternal life, of the king after death, as beautifully expressed in the Pyramid Texts two hundred years later.[22] If this were the case, then the pyramids would be gigantic, physically real – but symbolic – manifestations of the eternal power of the king integrating a huge and impressive cosmic landscape (see Fig. 2). This once more would speak of a unified project that would allow the king to manifest himself, both during his life and after death, as conspicuous celestial aspects of kingship and the divine.

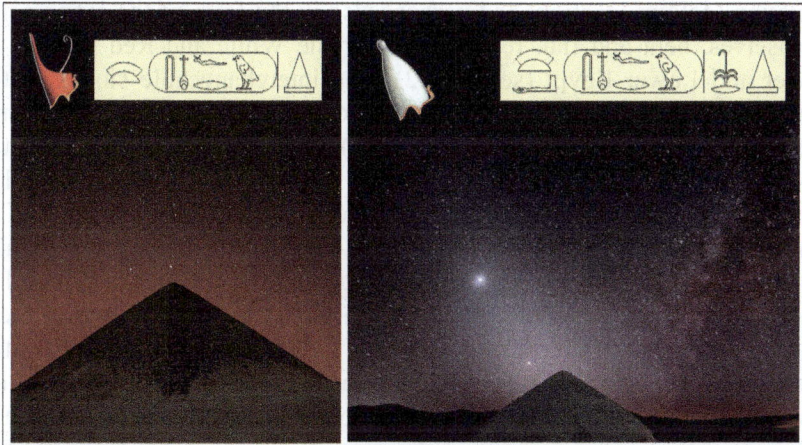

Fig. 2. Astronomy, architecture and symbolism at Dahshur: the Red (left) and Bent (right) pyramids as symbolic representation of the Red and White crowns, respectively, as petrified counterparts of relevant respective reddish and whitish celestial phenomena such as the aurora borealis or the zodiacal light (plus Venus). See the text for further details. Diagram by the authors, courtesy of the Multimedia Service of the IAC, on images courtesy of Graham Parkin and Daniel López. Adapted from J. A. Belmonte and G. Magli's paper, 'Astronomy, architecture and symbolism: the global project of Sneferu at Dahshur'; see that paper for a lengthy explanation of these arguments.[23]

---

[22] D. A. Cintron, 'A New Angle on Sneferu's Pyramids'. This paper is the unpublished outcome of a presentation at the 54th Annual Meeting of the American Research Centre of Egypt, in Cairo in 2003. For the Palermo Stone, see T.A.H. Wilkinson, *Royal Annals of Ancient Egypt: The Palermo Stone and its Associated Fragments* (New York and London: Routledge, 2000), pp. 142–45.

[23] Belmonte and Magli, 'Astronomy, Architecture and Symbolism', pp. 173–205.

The story could be summarized as follows: in year 8 or 9 of his reign, and for unknown reasons, King Sneferu decided to abandon his necropolis at Meidum, founding a new residential area at the limits between Upper and Lower Egypt. Next to it, a new, unique project of a dual nature was developed for the Pharaoh's afterlife, but also as a true image of his overwhelming power in life. This was conceived as a funerary complex integrated by two gigantic pyramids with their respective mortuary temples, a satellite pyramid (perhaps for the Ka of the king) and (at least one) 'Valley' temple.

The two – Bent and Red – pyramids would demonstrate Sneferu's sovereign power as dual-king of Upper and Lower Egypt by symbolic imitation (colour, location, and perhaps also in the form of the monuments themselves, bent and flat, respectively) of the White and Red Crowns of Upper and Lower Egypt (see Fig. 2). The interchangeable slopes of the pyramid were well defined with a peculiar astronomical symbolism (summer solstice and New Year's Eve) that could be related to the invention and further development of the Egyptian civil calendar. Astronomical alignments would also be included accordingly.[24] The pyramids were almost perfectly orientated towards the north, to the realm of the imperishable stars, and the access corridors were built with such a slope that they would facilitate the ascent of the king to the northern skies.[25] Besides, the pyramids themselves could be associated to celestial phenomena concomitant with the manifestation of king's power in his afterlife, notably either the zodiacal light or Venus (or both) for the White Crown and the aurora or light of dawn for the Red Crown (Fig. 2). The pyramids might then be considered as petrified light. This already existing astronomical symbolism would be put into writing two hundred years later in the 5th and 6th Dynasty Pyramid Texts.[26]

Sneferu presumably died in his 31st year of his reign, leaving the structures associated with the Red pyramid unfinished, notably his mortuary temple, which was quickly finished in mud-brick by his elder

---

[24] As largely discussed in Belmonte and Magli, 'Astronomy, Architecture and Symbolism', pp. 173–205.

[25] K. Spence, 'Ancient Egyptian Chronology and the Astronomical Orientation of the Pyramids', *Nature* 408 (2000): pp. 320–24; J.A. Belmonte, 'On the Orientation of the Old Kingdom Pyramids', *Archaeoastronomy* 26 (2001): pp. S1–20; E. Nell and C.L.N. Ruggles, 'The Orientation of the Giza Pyramids and Associated Structures', *Journal for the History of Astronomy* 45 (2014): pp. 308–69.

[26] For a complete discussion on these topics, see Belmonte and Magli, "Astronomy, Architecture and Symbolism', pp. 173–205.

surviving son and successor Khnum Khufu (Cheops), who would also develop his own very particular skyscape.

Akhet Khufu: an exceptional project

Early in the morning on 21 March 2005, the Egyptian researcher Mosalam Shaltout, Principal Investigator of the Archaeoastronomy Mission, and the author went to the Giza plateau. The intention was to watch the light and shadow effect produced at sunrise on the day of the equinox over the pyramids in an attempt to verify several hypotheses relating to this particular astronomical event, including the phenomenon of the flash of light on the north and south faces of the Great Pyramid.[27] After this and other efforts, we are now convinced that this phenomenon is extremely difficult to observe in the very moment of the rising sun, because the intensity of sunlight at that moment is too tenuous to produce the necessary shadows, sufficiently contrasted, to check the phenomenon. However, this might not have been the case in ancient times when the Tura limestone casing was on site.

Indeed, we were able to observe many other astronomical phenomena, such as sunrise just in front of the Sphinx, confirming the idea that this unique monument is aligned more or less to the sun rising on the 'Equinox' (see Fig. 3). However, this fact might simply reflect a general cardinal orientation of the huge pyramid complex of the 4th Dynasty at Giza, including the pyramids of Khufu and Khaefre (the Cheops and Kefren of classical sources) and their associated temples, with the north, as realm of the imperishable stars, as the most important reference source.[28] The Sphinx certainly is another part of the puzzle, although a very singular one, as beautifully illustrated in Figure 3. It is, with almost absolute certainty, a work of the 4th Dynasty, although there are strong debates if its construction must be attributed to Khufu, his youngest son, Khaefre, or even to his brother, and successor of Khufu, Djedefre.

---

[27] Belmonte, *Pirámides, templos y estrellas*, Figures 19 and 5.15.

[28] M. Shaltout, J. A. Belmonte and M. Fekri, 'On the Orientation of Ancient Egyptian Temples: (3) Key Points in Lower Egypt and Siwa Oasis', *Journal for the History of Astronomy* 38 (2007): Vol. 2, pp. 141–60 (Part I) and Vol. 4, pp. 413–442 (Part II); see also Belmonte, 'On the Orientation of the Old Kingdom Pyramids', pp. S1–20; and Belmonte, *Pirámides, templos y estrellas*, pp. 169–214.

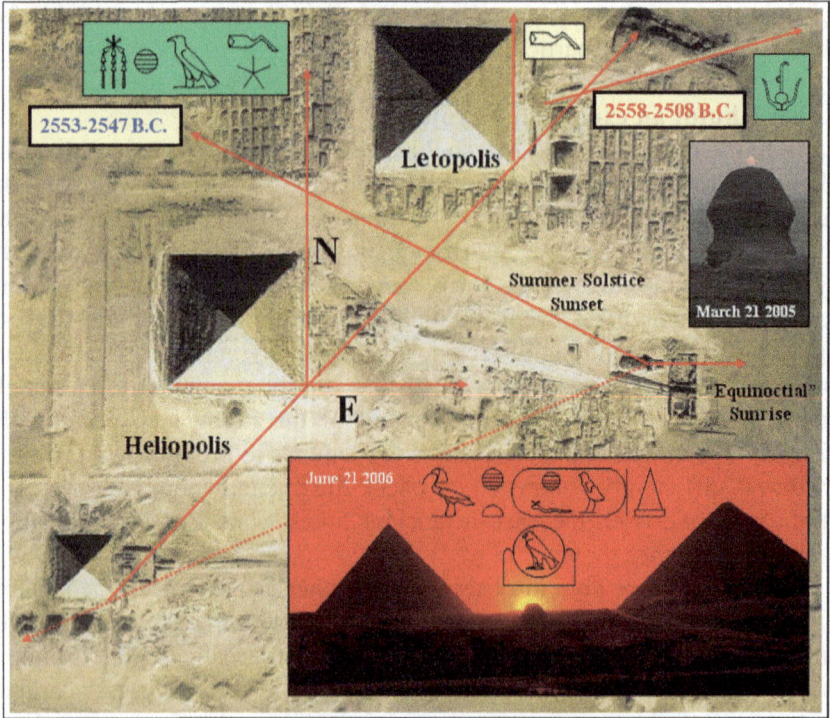

Fig. 3. Diagram showing the astronomical and topographic relationships between the different monuments erected on the Giza plateau – including the Sphinx and the pyramids – and certain elements of the sky and nearby geography. The relationship with the cities of Letopolis and Heliopolis was already proposed by Goyon in the 1970s.[38] However, this diagram also relates the original orientation of the pyramids to due north – based on observations of Meskhetyu – to the similar name of the province that had Letopolis as capital, the Bull's Foreleg. The astronomical connection of the Sphinx with dawn on the Equinox, and the emphasis on the alignment of Khufu's causeway to sunrise at Wepet Renpet (New Year's Eve) during his reign are also noticeable. The lower-right image illustrates the Sphinx of Giza and its alleged relationship to the two main pyramids. This image could be a reflection of the process of solarisation of the King during the reign of Khufu (ca. 2550 BCE), perhaps expressed in the name of the pyramid complex, Akhet Khufu, 'The Horizon of Cheops'. Later on, this image, in combination with the June solstice phenomenon, could have inspired the identification of the Sphinx with the God Hor-em-akhet, 'Horus at the horizon', during the New Kingdom. Diagram and images by the author.

The majority of Egyptologists support the identification of the head of the Sphinx as a portrait of Khaefre.[29] However, there is a dissident group, led by the prestigious German Egyptologist Richard Stadelmann, who believes that the Sphinx represented Khufu, regardless of if the portrait was sculpted during his reign or that of any of his two sons and successors, Djedefre and Khaefre.[30] Hence, the problem is not trivial.

One year later, on 21 June 2006, our group, the author included, returned to Giza to observe another – we guessed spectacular – astronomical hierophany (see Fig. 3). On this occasion, the idea was to look at the sunset on the summer solstice behind the Sphinx, from a position where the statue would be nestled by the two gigantic pyramids dominating the plateau. During the New Kingdom, the Sphinx was known as Hor-em-akhet, 'Horus at the horizon', and the Egyptologists Richard Wilkinson and Mark Lehner had independently suggested that this name would have been inspired by the similarity between the position of the gigantic sculpture in the middle of the two pyramids and the hieroglyphic term for Hor-em-akhet (Fig. 3):[31]

The phenomenon that was observed in Giza confirmed this suggestion by adding to the image proposed by Wilkinson and Lehner, the brightness of the sun-disk exactly behind the head of the Sphinx. Our intuitive idea, reinforced from this day on, is that this phenomenology could be deliberate and that the triad formed by the pyramid of Kefren, the Sphinx and the pyramid of Cheops was partly designed with this objective in mind.[32]

In conversations over the years with Giulio Magli, this Italian archaeoastronomer has defended the idea that there is no contemporary evidence that the pyramid of Khaefre was actually built by this Pharaoh.[33]

---

[29] Lehner, *The Complete Pyramids*, pp. 130–31; see also Z. A. Hawass, *Mountains of the Pharaohs* (New York: Doubleday, 2006), and references therein.

[30] R. Stadelmann, 'Las pirámides de la IV dinastía', in *Tesoros de las pirámides*, edited by Z. A. Hawass (Barcelona: Librería Universitaria, 2003), pp. 112–37.

[31] R. H. Wilkinson, *Symbol and Magic in Ancient Egypt* (London: Thames & Hudson, 1994). M. Lehner, 'Giza, a Contextual Approach to the Pyramids', *Archive für Orientforschung* 32 (1985): pp. 139–59.

[32] First proposed in Shaltout, Belmonte and Fekri, 'On the Orientation of Ancient Egyptian temples', Vol. 2, pp. 141–60 (Part I) and Vol. 4, pp. 413–42 (Part II); see also Belmonte and Magli, 'Astronomy, Architecture and symbolism', pp. 173–205, Figure 9.

[33] As most recently reflected in Magli, *Architecture, Astronomy and Sacred Landscape in Ancient Egypt*, pp. 83–89.

Colin Reader has even argued that works in part of Khaefre's causeway and other areas near the Sphinx began earlier than Khufu started to work on the place.[34] However, there is clear evidence that the Egyptians of the following dynasties had no doubt that the second Giza pyramid has to be assigned to Khaefre (e.g. the friezes in the tomb of Qar, priest of the Giza necropolis during the reign of Pepi I).

The story would in this case be as follows. Since his father Sneferu had constructed a huge dual project, Khufu (ca. 2550 BCE) might have decided that he could not be less than his predecessor and decided to plan another dual project but on an even larger scale. This unitary project would be integrated by two huge pyramids (larger than any of his father's ones), a huge statue of himself as a twin of the sun god (the Sphinx), associated temples and causeways, and a huge cemetery for the members of the royal family (the first occasion of a square grid in monumental construction). The new pyramid complex would receive the name of ⌦, Akhet Khufu, 'the Horizon of Cheops'(see Fig. 3). The idea of a unique project for the two largest pyramids of Giza was independently developed in the mid-2000s.[35] Actually, the possibility that both father and son imagined their burial monuments on such a gigantic scale reinforces each other and offers further support for the hypothesis of a common dual project in both cases.

A substantial difference would be that Sneferu had the time to complete, or nearly complete, his project while his son Khufu possibly left his unfinished after a long but shorter reign of at least 27 years. This would allow the idea that his younger son and second successor, Khaefre, was the one in charge of accomplishing the gigantic plan by simply 'usurping' nearly half of his father's project and assuming it as his own burial complex under the name of Ur Khaefre, 'Khefren is a great (of heaven)'. The name Akhet Khufu would since then have applied only to the Great Pyramid.

Such a huge dual construction project would have formed part of a large scale process of solarisation of the King rushed through during the 4th

---

[34] C. Reader, 'Giza before the Fourth Dynasty', *Journal of the Ancient Chronology Forum* 9 (2002): pp. 1–5.

[35] See Shaltout, Belmonte and Fekri, 'On the Orientation of Ancient Egyptian Temples', Vol. 2, pp. 141–60 (Part I) and Vol. 4, pp. 413–42 (Part II); and G. Magli, 'Akhet Khufu: Archaeoastronomical Hints at a Common Project of the Two Main Pyramids of Giza', *Nexus Network Journal* 11 (2008): pp. 35–50, plus G. Magli, 'The Cosmic Landscape in the Age of the Pyramids', *Journal of Cosmology* 9 (2010): pp. 3132–44.

Dynasty and begun by Khufu's grandfather, Huni, the first King of Egypt to write his name in a cartouche, continued, as we have discussed, by his father Sneferu, and which would reach its climax with Khufu, who would even have identified himself with the sun god Re.[36] As a likely consequence of this process, his son Djedefre would be the first King in Egyptian history to use the epithet 'Son of Re', a common term of the Pharaohs' titles thereafter. Therefore, the name Hor-em-akhet for the Sphinx would not have been more than a subsequent reinterpretation of the New Kingdom of something already floating in the atmosphere for many generations (see Fig. 3). Besides, the pyramid complex was orientated to the North, precisely, possibly to the simultaneous transit of Phecda and Megrez, two stars of Meskhetyu.[37] In this sense, the complex of Akhet Khufu would have aligned in the interval between 2559–2541 BCE (Fig.3).

But in Giza, not only astronomy but also the local topography played a significant role. Already in the 1960s, Georges Goyon proposed that the plan of the three great pyramids of Giza suggested a relationship with the cities of Letopolis and Heliopolis (Fig. 3).[38] The line that connects the southeastern corners of the three pyramids, with a slightly lower than 45° azimuth, was pointing toward the sacred city of the Sun God, while the direction towards the north led directly to Letopolis. In addition, the distance between Giza and Letopolis, and the latter city and Heliopolis, was almost 30,000 royal cubits (or 100 *stadia*). This suggestive finding had as a corollary that the respective sites of Giza, Letopolis and Heliopolis might have been selected deliberately. A curious fact is that Letopolis was the capital of the Egyptian province, or *nomo*, that was to the north of Giza, and whose emblem was a bull foreleg, the same used as a determinative for the name of the constellation of Meskhetyu (Fig. 3). Was there a connection between the celestial and terrestrial realities, as in the terms 'arctic' or 'septentrional' (both connected to the asterism of the Plough) in many European languages? If this was the case, at Giza we would be facing a Cosmic Landscape of gigantic scale.[39]

---

[36] Hawass, *Mountains of the Pharaohs*.

[37] As largely discussed in Belmonte, 'On the Orientation of the Old Kingdom Pyramids', pp. S1–20.

[38] G. Goyon, 'Nouvelle observations relatives à l'orientation de la pyramide de Khéops', *Revue d'Égyptologie* 22 (1970): pl. 7.

[39] As argued in Belmonte, *Pirámides, templos y estrellas*, pp. 217–22.

The siblings of the Sun God: temples, pyramids and texts

Firstly Shepseskaf, Khufu's grand-grandson, and subsequently his successor Userkaf (ca. 2470 BCE) left the family cemeteries to the north of Memphis to build their pyramidal complex close to the capital, at Saqqara. Shepseskaf chose a monument in the form of gigantic *mastaba* that seemed to connect with the pyramids of the founder of his dynasty, Sneferu, in Dashur; Userkaf manufactured a small pyramid in the shadow of the one of his glorious ancestors, Netjerkhet. However, this last King also built at Abu Ghurob, in the area of Abusir, the first in a series of solar temples, two of which (his and Niuserre's) have been archaeologically uncovered. His successors, the five 'solar' Kings of the 5th Dynasty (particularly Sahure, Neferirkare, Neferefre and Niuserre, and perhaps also Shepseskare) chose Abusir as their family necropolis (see Fig. 4). According to Stuart Jeffreys, the location of the dynastic temple complexes devoted to the solar god Re a bit north of the pyramid complexes of the kings − and in his opinion the reason for why the temples themselves were built − would have been deliberately selected with the objective in mind of remaining in view of Heliopolis, the sacred city of the Sun; a visibility that would have been lost not only from Saqqara but also from the pyramid complex of Abusir, further north (see Fig. 4).[40]

However, in the author's opinion there is a complete series of astronomical and topography connections between the pyramids fields of the 4th and 5th Dynasties (and indeed later of the 6th Dynasty), the solar temples of Abu Ghurob and the sacred cities of the capital area as shown in Figure 4.[41]

---

[40] D. Jeffreys, 'The Topography of Heliopolis and Memphis: Some Cognitive Aspects', in *Stationen. Beitrage zur Kulturgeschichte Ägyptens*, ed. Rainer Stadelmann (Mainz: von Zabern, 1998), pp. 63–71.

[41] See Figure 4. This is analysed in detail in Belmonte, Shaltout and Fekri, 'Astronomy, Landscape and Symbolism', pp. 213–84; and Belmonte, *Pirámides, templos y estrellas*, pp. 228–32.

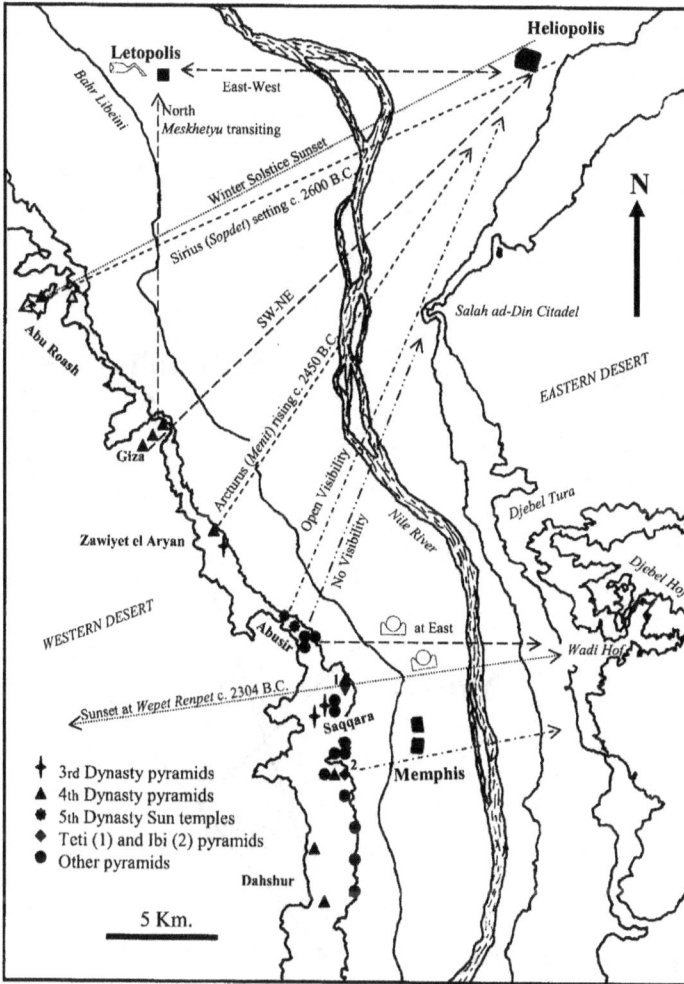

Fig. 4. Astronomy and landscape in the fields of pyramids of the Old Kingdom in the area of Cairo. The diagram shows some suggestive topographic and astronomical alignments. According to the subsequent analysis, the location of the pyramids of Abu Rawash, Giza and Abusir, and the solar temples at Abu Ghurob, could well have been deliberately chosen. The chart emphasizes the importance of Heliopolis, the city of the Sun God. Adapted from Belmonte et al. (2009).[42]

---

[42] Belmonte, Shaltout, and Fekri, 'On the Orientation of Ancient Egyptian Temples'.

Besides, a highlight of Magli's work is the idea of several elaborate dynastic landscapes, including most suggestive topographic and astronomical alignments, notably what he has termed 'symbolic invisibility' (see Fig. 5).[43] One advantage of this hypothesis is that it includes a prediction. The author forecasts where a few 'lost' pyramids of the Old Kingdom – notably Userkare's – might be located. If he were right on that, it would be a tremendous success and a confirmation of his ideas, which the author has found, although hypothetical, quite well grounded.

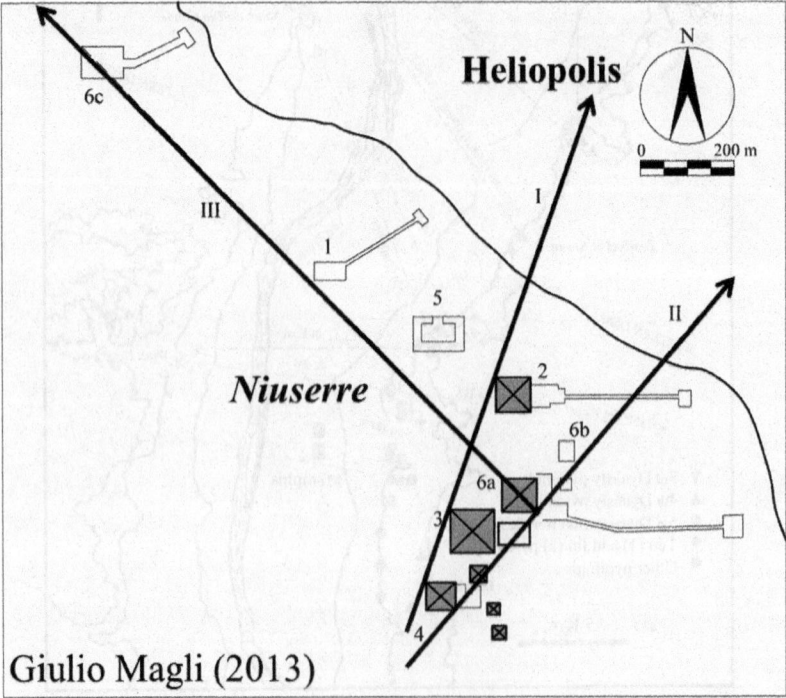

Fig. 5. The 'dynastic' diagonal of the 5th Dynasty (I) in the earlier pyramids of Abusir, pointing to Heliopolis. Niuserre's pyramid (6a) and solar temple (6c) were installed later following related but alternative patterns (II and III). Diagram adapted from Magli (2013).[44]

---

[43] Magli, *Architecture, Astronomy and Sacred Landscape in Ancient Egypt*, pp. 119–43.

[44] Magli, *Architecture, Astronomy and Sacred Landscape in Ancient Egypt*.

The pyramid complexes of the 5th Dynasty had an average orientation of half a degree (½°) to the southeast, if compared to those of their predecessors of the 4th Dynasty. This could mean that the usual method of orientation, which had been used for several generations since the reign of Sneferu, began to be imprecise, as might be suggested by the potential displacement in azimuth due to the precession of the equinoxes, when using the simultaneous transit of a couple of stars. Hence a new alternative method needed to be developed. Perhaps, due to the process of solarisation of the monarchy and the wandering nature of the calendar year with no leap-years, there also was a change of paradigm in the cult, reflected in a change of methodology in the form of aligning the sacred monuments, notably the pyramids; but this is difficult to ascertain.

However, this change could also shed light on the mysterious sudden appearance of the Pyramid Texts in the burial chambers of all the royal pyramids after the reign of Wenis, lasting until the end of the Old Kingdom. Perhaps, due to precession and the wandering nature of the calendar year, at the end of the 5th Dynasty there was a loss of confidence in the classical procedures of orientation that had so well permitted the King to perform his postmortem journey to the celestial realms of the 'imperishable' stars and the solar bark. For this reason, the guide to the afterlife was now written within the pyramid chambers for the exclusive use of the deceased kings (and queens).

The solstice, the New Year and the Nile: the paradigm of Ipet Sut
The temple of the god Amun-Re at Karnak, or Ipet Sut in the ancient Egyptian sources, shows an impressive axis of symmetry which can certainly be interpreted in a context where astronomy combines with religion, history and landscape to produce one of the most sacred traditional spots on Earth. The combination of the local course of the Nile, a solstitial orientation, the wandering aspect of the civil calendar and the nature of the deity worshipped in the temple might be considered as a paradigm of witnesses for the correct interpretation of the complex.

This magnificent religious complex located in ancient Thebes (modern Luxor) could have formed part of a relevant chapter in the history of archaeoastronomy. At the end of the nineteenth century, Sir Norman Lockyer argued that the main structure of the complex, the temple of Amun, would have been orientated towards sunset at the summer solstice, as the alignment of the main axis suggested.[45] However, when he asked for this hypothesis to be checked on site, he learnt that the hills of Western Thebes precluded such an alignment (see Fig. 6), and that the light of the setting sun never actually reached the interior chambers of the temple, unless the building had been constructed 56 centuries before, i.e. around 3600 BCE (change of declination due to the variability of the ecliptic).

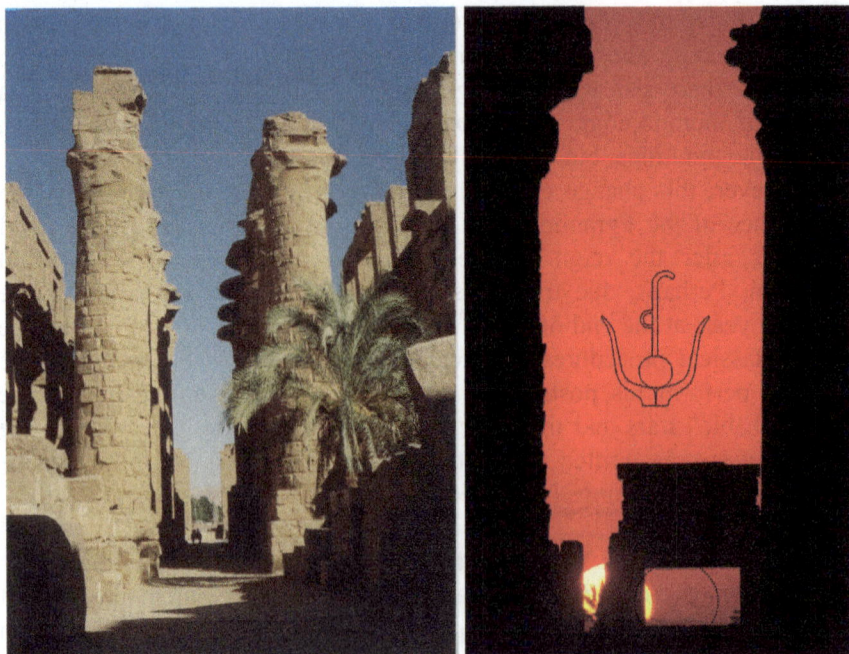

Fig. 6. The main axis of the Temple of Karnak, directed to the west (left). The hills of Thebes, located on the horizon, prevent observation of sunset at the summer solstice in the axis of the temple as could have been anticipated by studying the master plan of the monument. This fact leads to the suggestion that the opposite direction (towards sunrise of the winter solstice) is the one which could be relevant on the site, as demonstrated by sunrise (right) at the winter solstice in December 2006 in the main axis of the Temple of Karnak, as observed from the old pier of the sanctuary. The phenomenon would have been much more precise 4000 years ago when the temple was first erected during the Middle Kingdom. Images by the author. Adapted from Belmonte (2012).[46]

---

[45] J. N. Lockyer, *The Dawn of Astronomy* (New York: Dover Publications, 1993), new edition.
[46] Belmonte, *Pirámides, templos y estrellas.*

At his time, this date did sound problematic but still reasonable for the working chronology of the epoch. However, when the old chronology was abandoned and a new one proposed at the turn of the nineteenth century – with a reduction of a complete Sothic cycle, i.e., a shortening of 1460 years – his hypothesis was severely questioned. Apparently, although Lockyer was still alive and in his seventies, he never made the necessary effort to accommodate his proposals to the new chronology. Consequently, the potential solstitial alignment of Ipet Sut was forgotten for three quarters of a century.

Six decades later, a new argument was established, namely that the inscriptions of the walls of the complex supported the idea that, although the main temple entrance was open to the west and to the river, the temple was somehow connected to the east and especially to sunrise.[47] These ideas were later explored by Gerald Hawkins, who first reported on the winter solstice alignment of the 19th Dynasty temple of Re-Horakhty (included within Amun's precinct), but particularly called attention to the so-called 'high-room' of the sun, a sanctuary associated with the Hall of Festivals built in the complex by Thutmose III.[48]

A decade ago, it was demonstrated via archaeological excavations that the complex of the Amun temple at Karnak was surrounded during the Middle Kingdom by a village organized through a hyppodamian network whose main axis was that of the *dromos* connecting the temples of Mut and Amon.[49] This was also the orientation of the axis of Mut temple. However, the east-west axis of this network diverged by more than 7° to the main axis of the Amun-Re temple (109¼° vs. 116¾°, respectively). This fact suggests that the orientation of this temple was deliberately chosen and not at all restricted by local urban necessities. This is related to a couple of additional important facts.

---

[47] P. Barguet, 'Le rituel archaïque de fondation des temples de Medinet-Habou et de Louxor', *Revue d'Égyptologie* 9 (1952): pp. 1–22.

[48] G. S. Hawkins, 'Astroarchaeology: The Unwritten Evidence", in *Archaeoastronomy in Pre-Columbian America,* ed. A. Aveni (Austin: University of Texas Press, 1975), pp. 131–62; see also E. C. Krupp, 'Light in the Temples', in *Records in Stone: Papers in Memory of Alexander Thom*, ed. C. L. N. Ruggles (Cambridge: Cambridge University Press, 1988), pp. 473–99.

[49] J.-F. Carlotti, 'Considérations architecturales sur l'orientation, la composition et les proportions des structures du temple d'Amon-Rê á Karnak', in *Structure and Significance: Thoughts on Ancient Egyptian Architecture,* ed. P. Jánosi (Vienna: Verlag der Österreichischen Akademie der Wissenschaften, 2005), pp. 169–91.

It is important to notice that sunrise at the winter solstice may have had important mythological and/or calendrical implications. To be precise, because of the wandering of the civil calendar across the seasons, there have been two occasions after the creation of the calendar when New Year's Eve or I Akhet 1 fell at the moment of the winter solstice: the first one was in a four year period centred on 2004 BCE. This was a very interesting moment in Egyptian history. Mentuhotep II from Thebes had just re-unified the country and new buildings, on a large monumental scale, were constructed for the first time in the very south of the country, including Karnak, where the most ancient register – a polygonal column possibly of a door jamb – is dated from the reign of his father Antef III (ca. 2050 BCE). A few years later, the first well documented temple of Amun, indeed a new aspect of the solar god, was re-erected by Senuseret I in Karnak, on a large monumental scale and orientated to the rising of the sun at the winter solstice, precisely. This fact can hardly be ascribed to chance.

Besides, in the first half of the fifteenth century BCE something extraordinary happened in Egypt. A woman, the royal wife Hatshepsut, proclaimed herself dual king of Egypt. To do so, she had to proclaim that her father had been none other than the god Amun-Re himself, who had elected her for royalty, usurping the rights of the ruling king Thutmose III. At this time, the great temple of Ipet Sut had been standing for at least half a millennium since the time of the early Middle Kingdom, when it had been precisely orientated towards sunrise at the winter solstice.[50] However, the Middle Kingdom temple, and later enlargements by the first kings of the 18th Dynasty, faced west, towards the hill of Thebes. Instead, 'King' Hatshepsut built a new temple to *Amun-Re-who-hears-the-prayers* exactly on the same axis but open to the east, thus being the first structure at Karnak actually orientated towards sunrise at the winter solstice. Since this temple was in a court open to the public, we can only imagine the political revenues that such a divine hierophany would have accrued for Hatshepsut's interests.

On the death of Hatshepsut, the actual legitimate sovereign, her nephew Thutmosis III started his reign alone. Although it is not yet clear when the *dannatio memoriae* of Hatshepsut was performed, it is obvious that many monuments of the female 'king' were either usurped by the new King or

---

[50] L. Gabolde, 'La date de fondation du temple de Sésotris Ier et l'orientation e l'axe', in *Le Grand Château d'Amon de Sésostris Ier à Karnak* (París: de Boccard, 1998); see also L. Gabolde, 'Canope et les orientations nord-sud de Karnak étables par Thoutmosis III', *Revue d'Égyptologie* 1 (1999): pp. 278–82.

somehow lost prominence. This was the case of the temple of *Amun-Re-who-hears-the-prayers*. Thutmosis erected a new structure in front of it, thus preventing the illumination by the sun's rays of the statue of the Queen in close contact with god Amun. The main focus of this new structure was a single huge obelisk, the highest ever to be erected in Egypt – but later by his grandson Thutmosis IV– and which today adorns the Roman square of Saint John of Letran. This granite monolith was located exactly on the main axis of Ipet Sut. Afterwards, during the reign of Ramses II, the obelisk was surrounded by the structures of the new temple of Re-Horakhty, and the Hatshepsut's temple became sandwiched between two larger structures, the same situation in which we can see it today, making it difficult to imagine how it would have been when it was the first temple in the Ipet Sut complex facing the winter rising of 'her father Amun'.

In a different but convergent line of reason, Ipet Sut is located at the only site in the Nile Valley, above the first cataract, where the river flows in such a way that the average perpendicular direction to the water course is the solstitial line connecting winter solstice sunrise and summer solstice sunset. In fact, recent works by Luc Gabolde suggests that Karnak was founded on turtle-back elongated island, free of Nile floods, that was practically perpendicular to the solstitial phenomenon (see Fig. 7).[51] When the temple was enlarged, the Nile course adapted to this situation so that it started to flow in a course that was virtually perpendicular to the solstitial line, nearly the same situation as we can see it today. The author likes the idea, previously outlined by Edwin Krupp, that this natural accident could have been discovered by the Egyptians and may have helped to establish the sanctity of Thebes and, above all, the area of Karnak.[52] We would then be facing an extraordinary case of a combination of topography and astronomy; and indeed another paradigm of cosmic landscape.

The Star of Isis, the Nile and Meskhetyu: skyscaping at Denderah
We have so far dealt with sites which in some moment of Egyptian history became capitals of the country. We now singularly analyse a site where this was not the case but which has played a most important role in the history of cultural astronomy research in Egypt. We refer to Denderah.

---

[51] L. Gabolde, 'L'implantation du temple: contingencies religieuses et contraintes géomorphologigues', *Les Cahiers de Karnak* 14 (2013): pp. 3–12.
[52] E. C. Krupp, 'Egyptian Astronomy: A Tale of Temples, Traditions, Tombs', in *Archaeoastronomy and the Roots of Science,* AAAS Sym. 71 (1984): pp. 289–320.

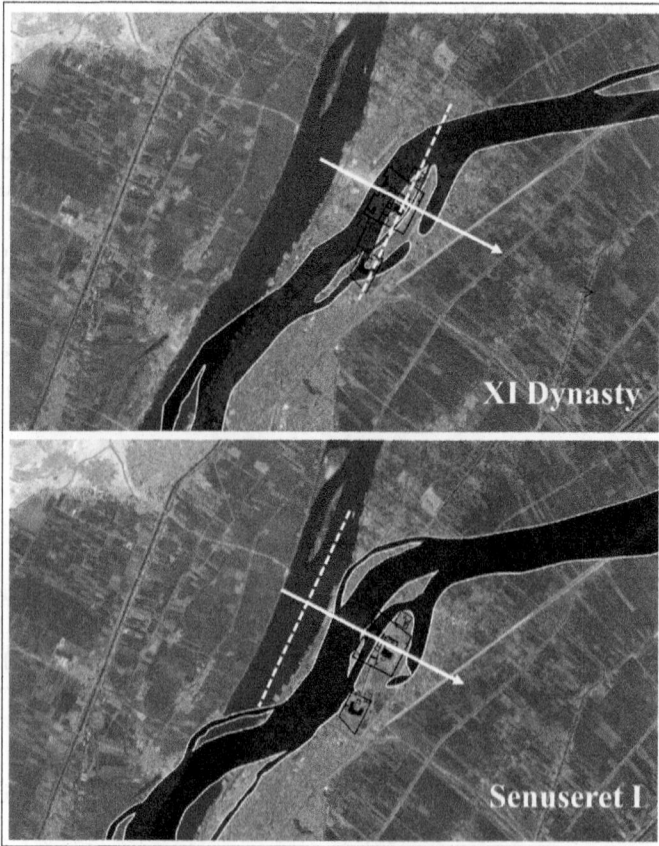

Fig. 7. Tentative topography of Karnak at the beginning of the Middle Kingdom. The temple of Amun, aligned to winter solstice sunrise (arrow) was presumably founded on a turtle-back island on the banks of the Nile. Later on, the temple was successively expanded to the west and the course of the Nile (dark grey) moved accordingly until reaching one that was fully perpendicular to the solstitial line. See the text for further discussions. Diagram by the author overlaid on a series of images courtesy of Luc Gabolde.

Throughout the 1990s, the team of the French Egyptologist Sylvie Cauville made a detailed study of the Temple of the goddess Hathor at Denderah. Studying texts based on certain epigraphic evidence, Cauville proposed that the axis of the main temple, dedicated to Hathor (see Fig. 8), could have been established on 16 July 54 BCE, during the reign of Ptolemy XII, father of Cleopatra VII. However, for the Temple of Isis, located behind the main temple, the situation was different.[53]

Fig. 8. Cosmic aspects of the Temple of the goddess Hathor at Denderah (a), oriented perpendicularly to the Nile and, at the same time, to the rising (or 'akh') of Meskhetyu on the date of construction as suggested by local inscriptions (b). In the astronomical diagrams of the temple, Hathor was assimilated with Isis-Sopdet (c). The main axis of the adjacent Temple of Isis, included in the complex, and the monumental East Gate of the *temenos* of the Temple of Hathor – where a processional way ended (d) – were facing the heliacal rising of Sirius (the brightest star of the sky, Egyptian Sopdet) in this area of the horizon during most of Egyptian history. Hence, Denderah temple might have been located in a place where the flow of the river would align with the direction in which the heliacal rising of Sirius, Herald of the Flooding, would be observed. In the Greco-Roman period, the perpendicular to this line also marked the 'akh' of Meskhetyu. The temples erected on the site had their orientations established accordingly. Diagram by the author.

[53] S. Cauville, E. Aubourg, P. Deleuzem and A. Lecler, 'Le temple d'Isis à Dendera', *Bulletin de la Société française d'Égyptologie* 123 (1992): pp. 31–48. Also relevant is S. Cauville, *Le Zodiaque d'Osiris* (Leuven: Peeters Pub & Booksellers, 1997).

This temple shows no less than three major axes: an oldest one, associated with earlier foundations in the reign of Nectanebo (30th Dynasty) and subsequent constructions of the reigns of Ptolemy VI and Ptolemy X; a processional axis directed toward a monumental gate on the Eastern wall of the *temenos* of the complex (see Fig. 8); and a festival hall dedicated to the birth of Isis, built in the time of Augustus. The former two varied 4° from one another while the third was at a 90° angle, so that the axis of the hall was parallel to the Temple of Hathor.

According to Cauville and her colleagues, the change of axis could be assigned as a change in orientation towards the rising of the star Sirius – most prominently associated to Hathor in the celestial diagrams of her temple (Fig. 8) – caused by precession. The oldest axis (with an orientation 111°11'), from the building of Nectanebo, should have maintained the prior orientation of a previous building, from which only some fragments were preserved and that would have been erected in the same location during the reign of Ramses II (ca. 1270 BCE). The new one, with an azimuth of about 108°, would point to the rising of Sirius in 54 BCE, when the axis of the new complex, including Hathor's temple, was established in its entirety.

This would mean that this new alignment was determined in accordance with the orientation of the main axis of the Temple of Hathor, as one might expect, but in a perpendicular direction. From the symbolic, sociological and even practical point of view, this solution seems reasonable, given that Isis, the owner of the smaller temple, is assimilated with Sopdet and therefore has Sirius as one of its major celestial manifestations. However, the inscriptions on the walls in the Temple of Hathor are crystal clear in this respect and what they say is that the astronomical object observed to establish its axis, and allegedly the plan of the entire complex, was the constellation of Meskhet(yu); in particular, towards the 'akh' of this constellation (Fig. 8).[54]

In 54 BCE, the rising of Alkaid (hUMa), the star located in the hoof of Meskhetyu's foreleg, and one of its most conspicuous stars, was visible at an azimuth of 18° and an angular height of about 2°, at the moment when all the constellation was fully visible after part of it had been hidden below the horizon, a celestial configuration that could be interpreted as the 'akh'

---

[54] For the inscriptions in detailed, see Belmonte, Molinero, and Miranda, "Unveiling Seshat', pp. 193–210.

(appearance) of the constellation.[55] Consequently, it is almost certain that, according to the Egyptians and their texts, the main Temple of Hathor at Denderah was oriented towards the rising of a singular star of Meskhetyu and not of Sopdet. Therefore, it seems obvious that at Denderah Temple Complex a combination of stellar alignments was produced in such a form that Sopdet and Meskhetyu would be the focal point of reference for different buildings at different epochs, such as the temple of Isis or the main temple of Hathor, respectively.

To this point, the opinion of a majority of specialists must be added: the idea that the Temple of Denderah is simply perpendicular to the Nile – which has a very peculiar course in that region of the Valley. The author is particularly fascinated with these series of apparent contradictions because it should inevitably lead to another of those wonderful examples where astronomy and the local geography combined to create another possible example of a cosmic landscape (see Fig. 8). The idea is that the temple of Denderah could have been chosen as a place of important worship for both Hathor and Isis – considered as two different manifestations of a same divine power at least from the New Kingdom – because the Nile runs almost parallel to the direction in which the heliacal rising of Sopdet was produced during a substantial period of Egyptian history. We could even dare to speculate with the suggestive metaphor that the waters of the flood would be announced by their herald star at the time of its earliest annually celestial manifestation: Peret Sopdet, precisely in this direction.

Later on, in the Hellenistic period, the obvious displacement of the position of the Sirius' rising due to precession could introduce certain doubts about the ability of the star as a precise reference for the orientation of the sanctuary. Hence, another procedure, already well known and used since ancient times (see Sections 1 to 3), would be selected, aligning the new temple with the stars of Meskhetyu. Therefore, we would be facing with a double stellar alignment that, in addition, would comply with the precept of being in agreement with the Nile; once more, astronomy and landscape in a close relationship.[56]

Epilogue
Pharaonic civilization temples in the Valley of the Nile were built in accordance with the local course of the river (either nearly perpendicular or

---

[55] As extensively discussed in Belmonte, Molinero, and Miranda, 'Unveiling Seshat', pp. 193–210.

[56] Belmonte, Shaltout, and Fekri, 'Astronomy, Landscape and Symbolism', pp. 213–84.

parallel to it) and, at the same time, obeyed certain astronomical patterns.[57] This means that the ancient Egyptians should have dealt with particular situations to meet both requirements. This problem was solved by the selection of orientation patterns that would comply with the constraints imposed by the Nile in different sites and, at the same time, obtain a conspicuous astronomical orientation that were in accordance with the religious and symbolic character of the temple. The examples we have seen in this paper are, in that sense, exceptional. More cases, such as Akehataten – the newly created city of Akhenaton at Tell el Amarna – the Late Period capital of Tanis, or the temple of the lunar god Thoth at Seikh Ibada across the river from Hermopolis, could have been also explored and would have shown relevant cases of a similar phenomenology.[58] Even cultures with strong Egyptian influences, such as the Kingdom of Kush in ancient Sudan, show sketches of the same character.[59]

In fact, the results presented in this essay review illustrate, with little margin of doubt, something that was not easy to imagine when our archaeoastronomical work in Egypt was begun more than one decade ago. This is that the sages of this exceptional culture scrutinized their environment, including the sky and their country's geographical framework, in their continuous fight against chaos and their quest for cosmic order; a reality that has been reflected in the close relationship encountered between astronomy, landscape and symbolism, in order to create actual cosmic landscapes.

Acknowledgments

I would like to express my gratitude to Nicholas Campion for inviting me to lecture on this fascinating topic during one of the Sophia Centre Conferences at Bath, England, in June 2014. This work has been mainly financed along the years under the framework of the projects P/310793 'Arqueoastronomía' of the IAC, and AYA2011-26759 'Orientatio ad Sidera III' of the Spanish MINECO.

---

[57] Belmonte, Shaltout and Fekri, 'Astronomy, landscape and symbolism', pp. 213–84, Figure 8.9.

[58] J. A. Belmonte, 'DNA, wine and eclipses: the Dakhamunzu affaire', *Anthropological Notebook* 19 (2013): pp. 419–41; and Belmonte, *Pirámides, templos y estrellas*, pp. 238–42 (for Tanis), pp. 246–47 (for Hermopolis).

[59] J. A. Belmonte, M. Fekri. Y. Abdel-Hadi, M. Shaltout, and A. C. González García, 'On the orientation of ancient Egyptian temples: (5) testing the theory in Middle Egypt and Sudan', *Journal for the History of Astronomy* 41 (2010): pp. 65–93.

Passages between Worlds:
Heaven, Earth and the Underworld in the Andean Cosmos

Kim Malville

And up the ladder of the earth I climbed
through the dreadful thickets of lost forests
to you, Machu Picchu.
High city of stepped stones
Home, finally for everything the earth
couldn't hide beneath its sleeping clothes.
In you, like two parallel lines,
the cradle of lightning and of man
rocked in a wind of spines.
Mother of stone, froth of condors.
Highest reef of mankind's dawn.

—The Heights of Machu Picchu
Pablo Neruda[1]

Abstract: Throughout the ancient Andean world there are architectural references to the three worlds and the means for traveling between them. The first people to visit Peru probably arrived by sea and ascent from one world to the next was a geographic imperative. The underworld is represented by the ocean itself, sunken circular plazas, caves sometimes with niches for mummies, and labyrinths. The heavens are represented by summits of pyramids, mounds, and high of peaks. Metaphorical passages between worlds are evidenced by monumental stairways and carved non-functional stairs associated with huacas. In the Inca empire, real passages involved climbing of some of the highest peaks of the Andes, the construction of ceremonial structures on their summits.

Introduction

The experience of many of the first people to reach the coast of Peru would have been that of ascent from the ocean to the mountains. This paper

---

[1] Pablo Neruda, *Alturas de Machu Picchu*, 1950; English translation by Lito Tejada-Flores, 1998.

Kim Malville, 'Passages between Worlds: Heaven, Earth, and the Underworld in the Andean Cosmos', *The Marriage of Heaven and Earth*, a special issue of *Culture and Cosmos*, Vol. 20, nos. 1 and 2, 2016, pp. 31–58.
www.CultureAndCosmos.org

proposes that the experience of climbing upward to the cordillera, combined with a shamanic symbolism of three worlds led to ritual ascents of truncated pyramids, the nearly ubiquitous non-functional carved stairways of the Incas, and, ultimately, the climbing of high Andean peaks. Water was an essential element of this theme of ascent. In Inca mythology, as part of the great hydrological cycle of the Andes, water was carried upward from the ocean to the snow covered mountains by the Mother Llama, *Yacana*. She appeared in the sky as a dark cloud constellation with the eyes of α and β Centauri , who drank water from the western ocean and carried it to the high mountains.[2] Origin stories of the Inca involve water either in Lake Titicaca or the ocean at Pacacamac or Ecuador,[3] which may be understood as a deep memory of ancestors arriving by sea and ascending inward. This paper suggests that the idea of ascent through three worlds was not only a geographic imperative, but it was also part of the shamanic tradition carried by early people who arrived in South America from Asia (Figure 1).

The First Americans
The first people to reach South America probably arrived by boat, coming out of Asia, keeping close to the coastline, navigating by sun, moon, and stars.[4] The earliest archaeological sites in South America are along the coast, hence migration along the Pacific Rim seems most likely. Studies of mitocondrial DNA have identified haplogroups that are spread along the entire west coast of the Americas, indicating rapid movement between 13.9 and 18.4 year BCE to reach southern Chile. It is noteworthy that the migration route along the western coast of South America identified by Bodner et al.[5] reaches the Araucanian-Mapuche people of Southern Chile. As evidence of the transfer of shamanic ritual traditions from Asia to South America. Eliade points to the similarities of the rituals of the Araucanian-Mapuche shamans and shamaness of southern Chile and those of Siberian

---

[2] Frank Solomon and George L. Urioste (translators), *The Huarochiri Manuscript* (Austin: The University of Texas Press, 1991).
[3] Gary Urton, *Inca Myths* (Austin: University of Texas Press, 1999).
[4] James E. Dixon, *Bones, Boats and Bison: Archaeology and the First Colonization of Western North America*. (Albuquerque: University of New Mexico Press,1990).
[5] Martin Bodner, et al., 'Rapid Coastal Spread of First Americans: Novel Insights from South America's Southern Cone Mitochondrial Genomes', *Genome Research* 22 (2012): 811–20.

shamans, including ascending to the sky by notched logs or trees and initiation in caves with animal heads.[6]

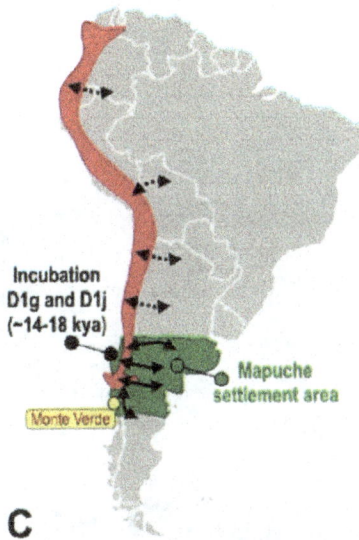

Fig. 1. Migration path along the coast of South America based upon studies of mitrocondrdial DNA haplocytes reaching Monte Verde and the Mapuche area. (after Bodener et al.[7] Cold Springs Harbor Press, Creative Commons License)

Soon after reaching Peru, the early visitors to Peru began to enter the high lands as revealed by recent discoveries[8] of an obsidian quarry, used by people who moved back and forth between the highlands and the coast. Ascent was thus a lived experience, given to them by the landscape they encountered. Near the quarry is Coropuna, at one time thought to be the highest mountain in the Andes. The Alca source, at an altitude of 4,800 m, Peru's largest highland obsidian deposit, was first used by Paleoindian

---

[6] Mircea Eliade, *Shamanism: Archaic Techniques of Ecstasy* (Princeton: Princeton University Press, 1964), p. 123.

[7] Bodner et al., 'Rapid Coastal Spread of First Americans'.

[8] Kurt Rademaker, Gregory Hodgins, Katherine Moore, Sonia Zarrillo, Christopher Miller, Gordon R. M. Bromley, Peter Leach, David A. Reid, Willy Yépez Álvarez, Daniel H. Sandweiss, 'Paleoindian Settlement of the High-altitude Peruvian Andes', *Science* 346 (2014): 466–69; Kurt Rademaker, 'Early human settlement of the high-altitude Pucuncho Basin, Peruvian Andes', Ph.D. thesis, (Orono: University of Maine, 2012).

foragers ~13,000 years ago living at the coastal site of Quebrada Jaguay. Even earlier dates for the peopling of South America have been obtained by Dillehay and colleagues[9] at Monte Verde, with plant material and bones dating to 18,500 years ago.

Fig. 2. Coropuna (photo by Edubucher)[10]

It is, indeed, not surprising that we encounter shamanism in the new world, brought by people moving along the Pacific rim. The legendary Peruvian archaeologist, Julio Tello,[11] who first excavated Chavín de Huantar, has provided us with a fascinating tale of a shaman traveling between the ocean and the high mountains.

---

[9] Tom D. Dillehay, Carlos Ocampo, José Saavedra, Andre Oliveira Sawakuchi, Rodrigo M. Vega, Mario Pino, Michael B. Collins, Linda Scott Cummings, Iván Arregui, Ximena, S. Villagran, 'New Archaeological Evidence for an Early Human Presence at Monte Verde', Chile, PLOS, November 18, 2015, *Science* 350 (2015): 898; *Todd A. Surovell, 'Simulating Coastal Migration in New World Colonization', Current Anthropology. Wenner-Gren Foundation for Anthropological Research 44, no. 4 (2003): 580–91.*
[10] Wikipedia Creative Commons License.
[11] Julio Tello, 'Wira Kocha', *Inca* 1 (1923), quoted in Hugh Thomson, *A Sacred Landscape: The Search for Ancient Peru* (New York: The Overlook Press, 2007) pp. 99–100.

Wari is the god of force. He constructed the irrigation dams and channels that remain from the prosperous agricultural past. When in former times the local shaman went to visit him in his mountain range, at the lake formed at the mountains foot by the melting ice, he would always take human blood mixed with maize flower as an offering.

When it does not rain in the mountainous areas and a bad year for agriculture is threatened, the shaman will descend to the coast,... and gather a small amount of water from the agitated and turbulent part of the sea; this water will then be carefully guarded in a canteen flask; through certain ceremonies the shaman identifies himself with the spirit of the water he is carrying up into the mountains. As he ascends he will from time to time let loose roars that imitate a feline beast; these cries are heard echoing around the Andes by the Indian population, and they are seen as a portent of rain. The shaman leaves some of the water in the wells that he passes, and on finally arriving at the lake, tips out the rest, little by little. A cloud is then said to come out of the canteen flask and darken the air, setting loose the storm which will inaugurate the rainy season. The shaman becomes the spirit of the water; he roars like a feline beast; he transforms himself into a puma; the sounds of the puma are the sounds of thunder; the water he carries is deposited in the wells to make them alive and sources of water themselves.

The story contains basic elements of Andean symbolism: the power of water; transformation of the shaman into a puma, ascent across the three words from the ocean to the mountains. As Thomson notes, 'the links between inland Peruvian sites and the sea were intense and binding'.[12]

Liminality

Liminality relates to passages, often transformative ones, from one realm to another, coming from the Latin word *līmen*, meaning a threshold. These passages may involve the ambiguity and disorientation that can occur when one crosses into a new and unfamiliar space or time. The idea of liminality was extensively developed by Turner in his discussion pilgrimage and the unsettling experience of traveling into unfamiliar landscapes.[13] While in the liminal state, human beings have a heightened awareness of their surroundings and are open to transformative suggestions from the environment or their companions. Liminality can involve places as well as

---

[12] Thomson, *A Sacred Landscape*, p. 99.

[13] Victor Turner, 'Liminality and Communitas', in *The Ritual Process: Structure and Anti-Structure* (New Brunswick: Aldine Transaction Press, 2008); Victor Turner, *Process, Performance, and Pilgrimage: A Study in Comparative symbology* (New Delhi: Concept, 1979).

experiences. Liminal places can range from springs, caves, shores, rivers, crossroads, bridges, and sacred spaces such as temples. In India, passages involving a 'crossing-over' are known as *tirthas*, the most famous of all is on the banks of the Ganga in Varanasi. Tirtha has multiple meanings; it is a place for passing from one side of a river to the other, a place for meeting gods who has passed from their realm into ours, and a place to pass out of this life.

Stairways, Spatial Similarity, and Fractal Interconnectedness

Fig. 3. Inca Steps with multiple scales: a. The Royal Mausoleum of Machu Picchu (Malville); b. River Intiwatana (Malville); c. Sayhuite The Rumihausi Stone with five scales of steps visible in the figure. (Steven Gullberg with permission).

Especially during the Inca Empire, steps carved into rock appear to have symbolized passages between worlds.[14] The rock outcrops or cliffs on which these steps were carved were perhaps themselves recognized as liminal, connecting the underworld from which they emerged and the present world that we occupy. In some cases, such as the steps cut into the northern side of Huayna Picchu at Machu Picchu, they lead upward from a cave probably containing mummies, the Temple of the Moon. The steps are frequently associated with caves, water, and springs, and may be places where our world opens up to the underworld. Passage through a double-jamb doorway at the base of the stairway may have signified entry into a liminal realm.

A fascinating feature of these stairs is their multiplicity of scales, such as the double scale in the Royal Mausoleum (Figure 3a), the River Intiwatana (Figure 3b) or the extraordinary *five* scales carved into the Third Stone of Sayhuite (Figure 3c). These may be examples of spatial similarity, i.e. ascent is conveyed in patterns, independent of size. This phenomena of multiplicity of pattern scale is also found in self-organized systems, is identified as scale invariance,[15] and is a characteristic of fractals and power law distributions.[16] In the Andes, the climbing of a pyramid, the experience of moving from the ocean to the cordillera of the Andes, or the ascent of a sacred mountain parallel the movement from one world to the next, all perhaps, contained in the meaning of carved non-functional stairs in huacas.

Analogism and Animism
There is a more complex and deep meaning to these carved steps than representing or symbolizing shamanic-like ascent or descent. These passageways between the worlds may have been living beings (Harvey: 'the world is full of persons, only a few of which are humans'.[17])

---

[14] Carolyn A. Dean, *A Culture of Stone* (Durham: Duke University Press, 2010).

[15] J. M. Malville, 'Complexity and Self-organization in Pilgrimage Systems', in *Pilgrimage: Sacred Landscapes and Self-Organized Complexity*, edited by J. M. Malville and B. N. Saraswati, (New Delhi: Indira Gandhi National Centre for the Arts, 2009).

[16] Per Bak, *How Nature Works: The Science of Self-Organized Criticality* (New York: Springer Verlag, 1996).

[17] Graham Harvey, *Animism: Respecting the Living World* (New York: Columbia University Press, 2006), p. 9.

Through the process of *camay,* water was an essential ingredient in animating the inanimate features of the high natural world, often bringing life to huacas.[18] While some of these beings may be like us, others may have been understood to be dangerous and unpredictable. For example, Pachamama produces earthquakes; apus may produce avalanches, malevolent winds, and fail to provide sufficient water for agriculture.[19] For ancient Andean people, the natural world may have been alive, but much of it may have been vast and complex beyond imagining.

The anthropologist Philippe Descola suggests that animism is but one of four ontologies involving living beings based upon their exterior and interior natures.[20]

	Interior similar to humans	Exterior different from humans
Exterior similar to humans	Totemism	Naturalism
Exterior different from humans	**Animism**	**Analogism**

Animists see many non-human objects as persons with self-awareness, thought, intentionality, and the ability to communicate. In their interior natures they are basically similar to humans, perhaps even humans in disguise. The living beings created by animists were most frequently found in gardens, jungles, and forests. Analogists, on the other hand, encounter beings that are fundamentally different from humans in both their interior and exterior natures, some would have been frightening huge and distant, such as the high peaks, the sun, and Milky Way. Descola suggests that such a cosmos may sometimes have seemed too incomprehensible to tolerate and that analogists create order by constructing analogies, bundling disparate things together. Such an invented bundle, he suggests, is the putative parallelism of macrocosm and microcosm. Andean people, living in a dangerous world of high peaks, earthquakes, floods, and avalanches,

---

[18] J. M. Malville, 'Animating the Inanimate: Camay and Astronomical Huacas of Peru', in *Cosmology Across Cultures*, edited by J. Alberto Rubiño-Martín, Juan Antonio Belmonte, Francisco Prada and Antxon Alberdi, (San Francisco: Astronomical Society of the Pacific, 2009), pp. 261–66.

[19] Inge Bolin, *Rituals of Respect: The Secret of Survival in the High Peruvian Andes* (Austin: University of Texas Press, 1998).

[20] Philippe Descola, *Beyond Nature and Culture* (Chicago: University of Chicago Press, 2013).

may have encountered a cosmos seemingly without primordial meaning. One approach to survive was to achieve reciprocal relations with these inscrutable and dangerous powers, perhaps through a compact involving offerings and requests.[21]

The disorienting chaos of the labyrinths of Chavín seem to reflect some of that terror of those inscrutable powers. Some sense of order and control over the world could also have been established by offerings at shrines as well as those ubiquitous carved stairways suggesting a strategy of survival through communication and contact with those powers. Paternosto describes the carved stairs as 'obsessive metaphoric representation of a communication.... between the world of the here and now...and the world beyond...'[22]

As suggested by Descola,[23] analogists encounter a cosmos consisting of a 'dizzying' atomism, of 'differences infinitely multiplied' with inscrutable purposes and meaning. Descola suggests that such a cosmos may have seemed too meaningless to tolerate. In response, disparate things were bundled together, such as a parallelism of macrocosm and microcosm. Perhaps this applies to Andean people, living in a dangerous world of high peaks, earthquakes, and avalanches. The disorienting chaos of the labyrinths of Chavín reflect some of that terror. Some order and control over the world could have been established by their non-functional ubiquitous carved stairways, suggestions of spatial similarity between the large and small, and hints of fractal interconnectedness. Their living huacas were not only extraordinarily powerful, but some were huge in scale such as the wall of Incamisana, the terraced pits of Moray, or the summit of Huayna Picchu. The mountains of the Andes were (and are) associated with living spirits, *apus*, who may have either resided in or actually have been the mountains themselves.[24] Some of these high mountains were ritually ascended and sacrifices were placed near their summits. These

---

[21] Laurence A. Kuzner, 'An Introduction to Andean Religious Ethnoarchaeology: Preliminary Results and Future Directions', in *Ethnoarchaeology of Andean South America*, edited by L. A. Kukznar, (Ann Arbor: International Monographs in Prehistory, 2001), pp. 38–66.

[22] César Paternosto, *The Stone and the Thread: Andean Roots of Abstract Art* (Austin: University of Texas Press, 1996), p. 71.

[23] Descola, *Beyond Nature and Culture*, p. 202.

[24] Carolyn Dean, 'Men Who Would Be Rocks: The Inca Wank'a', in *The Archaeology of Wak'as: Explorations of the Sacred in the Pre-Columbian Andes*, edited by T. L. Bray, (Boulder: University Press of Boulder, 2015), pp. 213–38.

apus of the highest world had great influence over humans; they may have
been perceived as alien giants who could be either dangerous or benign,
depending on how they were treated.

Chupacigarro/Caral

Stairways crossing the three worlds are found in the in the earliest
monumental structures of the Andes in the Supe Valley. The most famous
is the city of Caral[25] containing truncated pyramids, the major one is shown
in the figure, with a sunken circular plaza, some 20 meters in diameter with
a monumental stairway leading to the summit. The stairway would have
carried priests or shamans from the underworld to the highest heaven, the
summits of the sacred mountain where they could perform ritual and
ceremonies to bring water, snow, and rain to the land. It is not clear
whether the public viewed the rituals from the circular plaza, or if the plaza
was the lowest of the sacred realms, out of which shamans moved and
were transformed as they moved upward, as if becoming gods on the
summit of the sacred mountain.

The great plaza containing these pyramids is oriented along the line
from December solstice sunrise to December solstice sunset. The face of
the major pyramid is parallel to that line. There are at least ten other
temple-pyramids in the Supe valley, built in the period between 2500 BCE
and perhaps 3200 BCE.

Casma Valley

To the north of the Supe Valle in Norte Chico is the Casma Valley with
similar temples and elaborate mounds, built in the period between 0 BCE
and 2000 BCE. These immense manmade structures inform us about the
power of ritual ascent as well as the power of leaders to organize such
monumental constructions and their associated rituals. The major mound of
Sechín Alto was the largest structure in the western hemisphere at its time,
with a height of 44 meters and a volume of approximately 2,000,000 cubic
m.[26] The contemporaneous structure at Poverty Point in northeastern

---

[25] Solis R. Shady, 'America's First City? The Case of Late Archaic Caral', in
Andean Archaeology III: North and South, edited by W. H. Oibell and W. H.
Silvermann, (New York: Springer, 2006), pp. 28–66.

[26] J. M. Malville, 'Pre-Inca Astronomy in Peru', in *The Handbook of
Archaeoastronomy and Ethnoastonomy*, edited by C. Ruggles, (New York:
Springer 2015), pp. 795–806.

Fig. 4. Grand Pyramid of Caral (showing the monumental staircase leading from the sunken circular plaza). (Carlos Aranibar with permission)

Louisiana had a total volume of 240,000 cubic m. Monks Mound, the largest earthen structure of Cahokia has a volume of 622,000 cubic m, and the Pyramid of the Sun at Teotihuacan contains approximately 1,200,000 cubic m. The major axis of Sechín Alto containing a monumental stairway is oriented toward June solstice sunrise.

Chavín de Huatar
Chavín de Huantar[27] perhaps dominated the spirituality of the Casma valley; it was apparently a major pilgrimage destination and to reach it many people would have walked up the valley, over the pass of the Cordillera de Negra into the valley and across another pass, then downward into Chavín. This ceremonial probably had immense influence in the Andean world.

---

[27] Richard L. Berger, *Chavín and the Origins of Andean Civilizations* (London: Thames and Hudson, 1992); W. J. Conklin and J. Quilter, eds., *Chavín: Art, Architecture, and Culture* (Los Angeles: Cotsen Institute of Archaeology, 2008).

Fig. 5. Descent into the Labyrinth, Chavín, 1975. (Malville)

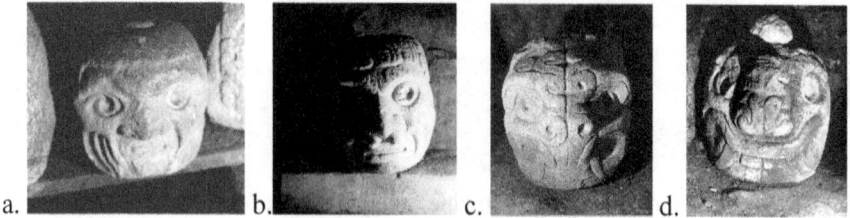

a.   b.   c.   d.

Fig. 6: Faces in the Labyrinth, images of the underworld. (Malville)

The labyrinths or galleries of Chavín appear to be miming the underworld.
The principal deity of Chavín may have been enshrined in the 4.5-meter-
high carved shaft of stone named the Lanzon by Tello, containing an image
of a fierce fanged deity, similar to some of the heads emerging from the
sides of the labyrinth. A slab over the top of the Lanzon could have been
removed and offerings poured onto it, or an oracle stationed above the
Lanzon could have spoken to those gathered below in the darkness. The
Lanzon may have been the axis mundi, separating the worlds, propping
them apart or a providing a passage between them.

The galleries are composed of narrow passageways, with sharp bends,
stairways, pits into which one might fall, and small chambers. The Gallery
of Offerings consists of a 24-meter-long passage with 19 dark rectangular
cells, perhaps for mummies or for terrified and disoriented celebrants.
Further evidence of the foreboding nature of the place was the discovery of
a female skull contained in a circle of 40 human teeth. Elsewhere in the

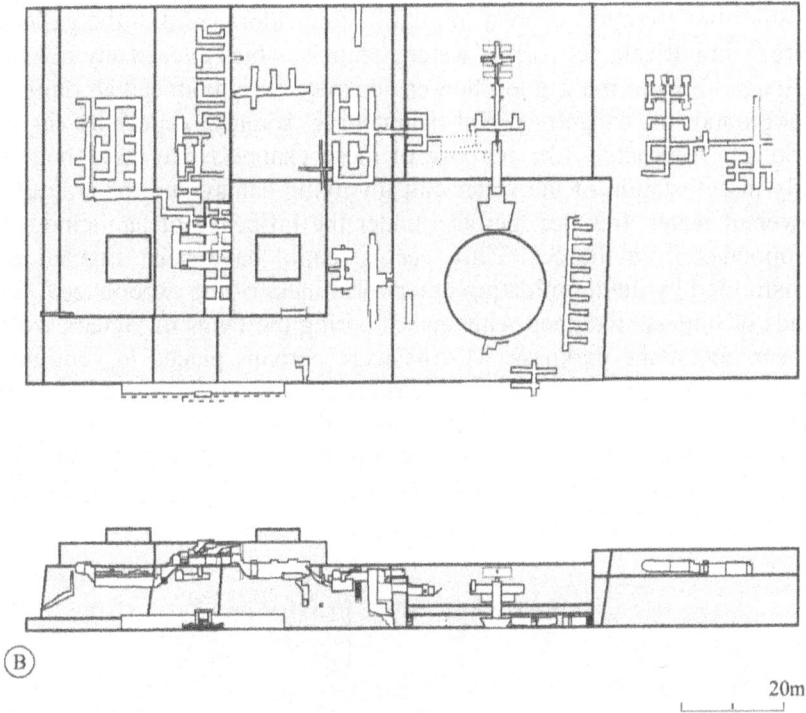

Fig. 7. Chavín and it's Galleries or Labyrinths. To the right (north) of the Circular Plaza is the Gallery of the Offerings. The many staircases are shown in the profile. (Kembel, Figure 4.2,[28] with permission)

Gallery were bones of over 20 individuals, which together with the female may be evidence of ritual cannibalism.[29]

---

[28] S. R. Kembel, 'The Architecture at the Monumental Center of Chavín de Huantar: Sequence, Transformations, and Chronology', in *Chavin: Art, Architecture, and Culture*, edited by William J. Conklin and Jeffrey Quilter, (Los Angeles: Cotsen Institute of Archaeology, 2008), p. 268.

[29] Tiffiny A. Tung, 'From Corporeality to Sanctity: Transforming Bodies into Trophy Heads in the Pre-Hispanic Andes', in *The Taking and Displaying of Human Body Parts at Trophies by Amerindians*, edited by Richard J. Chacon and David H. Hyde, (New York: Springer, 2007), pp. 481–504; Burger, *Chavin and the Origins of Andean Civilization*, pp.138–39.

Using the maze of passageways as disorienting venues, they constructed elaborate systems to manipulate light and sound. The sound of Strombus trumpets echoed through the corridors. Within the galleries there is an intricate network of water channels, which exceed any need for drainage.[30] Water roaring through canals may have produced an ominous background of dangerous and frightening sounds, especially in the darkness. Alternately, the purpose of these channels may have been an early manifestation of the water cult involving camay and the animating power of water. Initiates, perhaps under the influence of hallucinogenic compounds from the San Petro cactus, could have been stunned and transformed by the intimidating otherworldliness of the experience.[31] The heads of humans wrapped with snakes baring the fangs of jaguars would appear out of the darkness. Mirrors were perhaps placed in ventilation ducts to reflect the sun into the subterranean hallways and manipulate the darkness. A truly liminal experience would have been created in this intense sensory environment in which the normal world had been upended, coming as close to the underworld as one could imagine and to powers vast and dangerous.

Fig. 8. Perhaps a procession of Shamans at Chavin. Note the wings, suggesting shamanic flight. The first figure is holding a conch and the second what appears to be a spiny oyster, *Spondylus,* providing a connection to the ocean. (Rick, Figure 1.15,[32] with permission)

---

[30] D. A. Congeras and D. K. Keffer, 'Implications of the Fluvial History of the Wacheqsa River', *Geophys J. Royal Astronomical Society* 24 (2009): pp. 589–618.
[31] John W. Rick, 'Context, Construction, and Ritual in the Development of Authority at Chavin de Huantar', in Conklin and Quilter, eds., *Chavin: Art, Architecture, and Culture*, pp. 27–29.
[32] Rick, 'Context, Construction, and Ritual', p. 21.

Chankillo

Chankillo has become one of the most frequently described sites in the Casma Valley. It is a place of stairways: 25 stairs connect its thirteen towers. These towers are themselves ascending platforms, built successively higher such that, as one moves upward platform by platform, one moves to higher and higher worlds. The orientations of the platforms slowly rotate from a terrestrial world (north-south to alignment with the solstices, as if celebrants are moving from the terrestrial to the solar world. The stairways are steep, making for arduous passage, not uncommon amongst pilgrimage experiences. Only the highest tower has one stairway, on its north side, indicating that it was the final destination for upward ritual movement.[33]

Ollantaytambo

The Inca ruler, Pachacuti, is credited with establishing Cusco as the capital of the Inca Empire, embarking on a course of imperial expansion, and setting up the system of carved outcrops as huacas of the Cusco basin. Christie[34] has suggested that Pachacuti created the style of geometric carving of rocks for both geopolitical and ideological purposes. These rocks established the reach of the Empire and created huacas, which had power and wisdom. He established royal estates at Pisac, Ollantaytambo, Machu Picchu, and Vitcos, each of which contains elaborately carved stones. Ollantaytambo contains the hallmarks of Pachacuti's reign: a ceremonial center containing massive stonework, elaborate stone carvings, and the water shrine of Incamisana with water channels and elaborate fountains.[35] Cliffs on the western side of the Incamisana contain a myriad of carved nonfunctional stairways, waters channels, and niches (Figure 8). The nonfunctional stairs are extraordinary in their 'obsessive' abundance.[36] The cliff face may have been a massive huaca, animated and empowered by sunlight and the flowing water at its base.

---

[33] Shelia Porzorski and Thomas Pozorski, *Early Settlement and Subsistence in the Casma Valley, Peru* (Iowa City: Univeristy of Iowa Press), p. 99.

[34] Jessica J. Christie, 'Inka Iconographic Rocks', IFRAO Proceedings, *American Indian Art* 40 (Albuquerque: American Rock Art Research Association, 2013), pp. 111–24.

[35] K. R. Wright, A. M. Gibaja Oviedo, G. F. McEwan, R. D. W. Miksad, and R. M. Wright, *Incamisana: Engineering an Inca Water Temple* (Reston: American Society of Civil Engineers, 2016).

[36] Paternosto, *The Stone and the Thread*, p. 71.

Fig. 9. Ollantaytambo and the Water Temple of Incamasina. a. The Incamasina seen from across the valley, to the right of center between the terraces. (Carlos Aranibar with permission) b. Carved steps (Malville).

Moray

In the altiplano 35 km northwest of Cusco, as part of his estate of Chinchero, Topa Inca, the son of Pachacuti, fashioned three natural sinkholes into a remarkable set of terraced basins. A long-held interpretation of the basins as an agricultural research station is due to the work of John Earles.[37] His measurements of the soil temperatures on different terraces led him to propose that each terrace had a different microclimate and could therefore have been utilized for testing and modifying a variety of crops.

A major breakthrough in the understanding of Moray came from the hydrological analysis of the Wright Paleohydrological Institute.[38] Their work established that the basins could not have functioned as irrigated agricultural lands. Water channels were designed to feed water to drop structures near the southern end of the basins, but water would have had to flow uphill to reach the northern portions of the terraces. The terraces of Moray are landslide prone; by adding water for irrigation the landslide problem would be exacerbated. Finally, due to evaporative cooling on all terraces, there are no significant temperature differences across terraces. Because of the failure of the agricultural hypothesis, Wright and colleagues have concluded that the basins were primarily designed for ceremony and ritual. The shaped basins could have been 'inverted' huacas, animated by the water flowing into them. The basins may also have served as ushnus into which water offerings flowed to Pachamama. Ceremonies intended to encourage rainfall may have taken place when water was released from the reservoirs above the basins. A dramatic public ceremony may also have occurred on the days of the zenith sun, when terraces would cast no shadows and the light of the sun would pass directly into the earth and Pachamama. The stairways leading downward into the basins may have been used for ritual processions, metaphorically entering or emerging from the lowest world.

---

[37] J. Earls, *Planificación agrícola andina: Bases para un manejo cibernético de sistemas de andenes* (Lima: Univ. del Pacífico y COFIDE, 1989); J. Earls and I. Silverblatt, 'The instrumentación de la cosmología Inca in el sitio arqueológico de Moray',in Lechtman and Soldi (eds) *Runakunap Kawsayninkupaq Rurasqankunaqa: Lastecnologías en el mundo andino* (Mexico DF: Univ. Nac. Auton. de México, 1981) pp. 433–74.

[38] K. R. Wright, Ruth W. Wright, A. Valencia Zegarra, G. McEwan G, Moray, *Inca Engineering Mystery* (Reston: American Society of Civil Engineers Press, 2011).

Fig. 10. The Flying Stairs of Moray (Tore Lomsdalen with permission)

Machu Picchu

Perched on a narrow granite ridge above the encircling Urubamba River,
Machu Picchu embodies all the elements of passages between worlds and
ascent that this paper has been discussing. An important huaca of Machu
Picchu is the Temple of the Condor, which has a carved rock apparently
representing a condor, a symbol of the heavens. Behind the condor is a
cave with a non-functional stairway leading to a dark natural fissure.
(Figures 11 & 12) The cave is briefly illuminated by the rising sun on the
day of the anti-zenith sun.[39] This remarkable combination of bird, dark
staircase, and anti-zenith sun may have been intended as analogous to a
passage between the underworld and the sky.

---

[39] J. Westerman, 'Inti, the Condor and the Underworld: The Archaeoastronomical
Implications of the Newly Discovered caves at Machu Picchu, Peru', in *Current
Studies in Archaeoastronomy,* edited by J. W. Fountain and R. M. Sinclair,
(Durham: Carolina Academic Press, 2005), pp. 339–51.

Fig. 11. The Temple of the Condor (Malville)

Fig. 12. Stairs to the Underworld in the Temple of the Condor (Malville)

Framed in the major entrance to Machu Picchu is the peak of Huayna Picchu with two stairways up its southern and northern sides. The southern route, which starts in Machu Picchu, passes through a narrow cave near the summit. On the southern stairway, one crosses a large observing platform and then enters this narrow, dark tunnel (difficult if one is wearing a backpack) and climbs its stairs, moving, as it were, from the mundane world of Machu Picchu to the upper world of the heavens. The northern route starts at the Urubamba River and continues through the largest double-jamb doorway of Machu Picchu, near the cave of the Gran Caverna formerly known as the Temple of the Moon. This large cave contains beautifully formed double-jamb niches, perhaps places for mummies of important ancestors. It is possible that these mummies were carried on the backs of celebrants from the underworld of the cave to the sunrise on upper world of the summit of Huayna Picchu for special celebrations. Admittedly, this is a highly speculative suggestion, but it would have been entirely consistent with the public display of royal mummies that occurred in the plaza of Cusco on the important ritual days of the year.[40] This combination of stair, cave at the bottom, places for ceremony on the summits makes Huayna Picchu powerfully emblematic of passage between the lowest and the highest worlds.

Fig. 13. Temple of the Moon (Gulberg with permission)

[40] Brian Bauer, *Ancient Cuzco: Heartland of the Inca* (Austin: University of Texas Press, 2004), pp.159–84.

Fig. 14. The southern side of the summit of Huayna Picchu. The House of Three Windows is on the left-hand skyline. (Tore Lomsdalen with permission)

Fig. 15. The Tunnel Below the Summit of Huayna Picchu
(Kenneth Wright with permission)

Fig. 16. Stairs on the North Side of Huayna Picchu (Malville)

Fig. 17. The area of the Temple of the Moon, showing the double-jamb doorway
leading to the summit of Huayna Picchu. (Malville)

The Ultimate Passage: Ascending Sacred Mountains

Huayna Picchu is a smaller version of the great sacred mountains of Peru,
many of which pilgrims climbed. Another example of a smaller sacred
mountain is the conical hill of Sondor, which appears as an inversion of the
sculptured pits of Moray. It contains a stairway with two double-jamb
doorways, one at the bottom and one at the top, which leads upward to
sunrise on the morning of the zenith sun. On the mornings of the day the
sun reached the zenith, it may have been climbed by parties of pilgrims. A
huaca is at the summit.[41]

There are reports of the Spanish Chroniclers that the Incas made annual
pilgrimages and offerings to major mountain deities at special times of the
years.[42] If the peak was too difficult to ascent, offerings would be made

---

[41] M. Zawaski and J. McKim Malville, 'An Archaeoastronomical Survey of Major
Inca Sites in Peru', *Archaeoastronomy: The Journal of Astronomy in Culture* 21
(2010): pp. 20–38.
[42] Johan Reinhard, *The Ice Maiden: Inca Mummies, Mountain Gods, and Sacred
Sites in the Andes* (Washington DC: National Geographic Society, 2005); Johan
Reinhard and Maria Constanza Ceruti, *Inca Rituals and Sacred Mountains: A*

from a place in view of the mountain, such as an ushnu, into which water, chichi, or, perhaps, llama blood would be poured. Sometimes offerings were thrown toward the summit. There are more than 100 peaks above 17,000 feet that were ascended by the Inca, containing shrines or other evidence of ritual activity.

The Incas are, indeed, renowned for their marvelous architecture, skillful masonry, political organization and for their extensive system of roads. Perhaps the most remarkable achievement of their culture is the ascent and placing of structures on many of their highest peaks, the highest being Llullaillaco with an altitude of 22,110 feet. It is the seventh highest peak of the Americas, containing on its summit the world's highest archaeological site, which has been carefully excavated by Johan Reinhard, Maria Ceruti, and colleagues.[43] Reinhard has been the leader in locating and excavating many of these high archaeological sites, and it is to him that we turn for the definitive description of these remarkable summits, especially that of Llulliallaco.

The most important mountains were the sites of capacocha ceremonies, which involved human sacrifice, mostly boys and girls chosen for their beauty and perfection. These ceremonies on mountain summits may have been offerings to the sun, Inti, the weather god, Illapa and mountain deities. Offerings on the summits may have been attempts to get closer to the sun. Some of the sacrifices were intended as a marriage between the girl victim and the mountain god, with whom the girl was to live for eternity.

Capacocha pilgrimage to mountain summits could last weeks or months of travel, covering distances of 1000 km or more. These pilgrimages included priests, assistants, local inhabitants, the child to be sacrificed, and sometime his or her parents. The procession would stop at sacred places along the way to make offerings. When passing through the mountains the pilgrims would reportedly keep as silent as possible to avoid angering the mountain gods.[44] The priests leading the procession looked only straight ahead with their heads lowered. Sometimes blood of sacrificial llamas was carried to place as offerings at huacas along the route.

Study of the World's Highest Archaeological Sites. (Los Angeles: Cotsen Institute of Archaeology, 2010.)

[43] Reinhard and Ceruti, *Inca Rituals and Sacred Mountains.*

[44] Reinhard and Ceruti, *Inca Rituals and Sacred Mountains*, pp. 88–89.

Fig. 18. Inca rooms and the platform on the summit of Llullaillaco, March 15, 2016. (Graham Zimmerman with permission)

Llullaillaco is the seventh highest peak in the Americas, just 100 lower than Huascaran, the highest peak in Peru, lying on the border between Chile and Argentina, in the barren Atacama Desert. To reach it would have been an arduous journey for pilgrims, perhaps taking weeks or a month. Access to its summit is possible only five months of the year, between November to March, which includes December solstice, Capac Raymi, one of the major Inca festivals.

The pilgrimage trail to Llullaillaco contains several structures, which appear to be resting places for pilgrims, priests, and sacrificial victims. The largest of these way stations is a tambo at 17,000 feet, which could have housed 100 people, perhaps the majority of the pilgrimage party, who did not venture to the summit of the mountain. There are three, smaller, intermediate sites at 18,325', 20,669', and 21, 325'. The capachocha priests, victim, and assistants would probably have reached the summit on the afternoon of the day before ceremonies and spent the night in the small summit building, which consisted of two rooms which may have been roofed with grass mats. The burial platform, measuring 10m x 6m and 50 cm in some places, contained the bodies of three sacrificial victims, 15-year-old woman, 6-year-old girl, and a 7-year-old boy. The well-established trail, clearly visible in Figures 18 and 19, leading to the platform suggests there may have been more visitors to the summit than those associated with three capacocha ceremonies.

Fig. 19. Summit Platform on Llullaillaco (Graham Zimmerman with permission)

Because the burials in the ceremonial platform were in undisturbed condition when excavated, we have evidence of astronomical meaning in the capacocha ceremony. The young boy was approximately oriented along the short side of the rectangular platform, facing east-southeast. The long wall was measured by several observers to have an orientation of approximately 30° magnetic, with an uncertainty of several degrees due to the irregularities of the wall (see Figure 19) and the difficulties of measurement under the severe conditions on the summit.[45] That orientation corresponds to an azimuth for the short side of the platform of 118.7° true. The first gleam of the sun on December solstice 1500 CE on a flat horizon would have an azimuth of 116.6° based upon a refraction of .6°. Considering the additional uncertainties of unknown refraction at such an altitude, the exact orientation of the mummy, and the extreme difficulties of the Inca in constructing a platform on the summit, it is possible the boy was intended to be buried facing sunrise on a date close to December solstice, the date of Capac Raymi, when the mountain was climbable. This extraordinary ceremony involving an arduous and lengthy pilgrimage to the base of the mountain and the difficult ascent to the highest mountain yet climbed by humankind is paradigmatic exemplar of a liminal passage way between worlds, the best example we have in the Inca cosmos.

---

[45] Johan Reinhard, personal communication.

Concluding Remarks

Passages between worlds in Andean cultures started with the first visitors as they moved upward and downward as they traveled between their places of origin at the edge of the ocean to the bases of snow peaks such as Corupuna. The wider significance of such journeys between world may have been enhanced by traditions of shamanism that they brought with them from Asia.

Truncated pyramids in the Supe and Casma valleys and Chavín de Huantar, with monumental stairways leading to their summits, appear to provide surrogate mountains for ritual ascent between the worlds. These structures are frequently aligned along solar axes. The sunken circular plaza of Caral suggests the symbolism of movement from the lowest to the highest world. The labyrinthine galleries of Chavín provided powerful liminal experiences of the lowest of the three worlds. The multiple stairways of Chankillo continued the theme of ritual movement between worlds. Centuries later the non-functional stairs caved into Inca huacas repeated this theme in iconic form, with an additional hint of fractal-like scale invariance, i.e., scale didn't matter in the process of passage between worlds. Also, during the Inca empire, opportunities for *actual* ritual ascents to the upper worlds occurred at Huayna Picchu, Sondor, and high Andean peaks, culminating in the climbs to the summit of Llullillaco.

The Zenith Sun as an Organizing Principle of the Constructed Sacred Space and Calendrics of Central Mexico

Harold H. Green

Abstract: Speculations that the Mesoamerican 260-day *tonalpohualli* may have a direct relationship to the 260-day interval between solar zenith passages that uniquely occur within a narrow latitudinal band just south of 15° north latitude (the '260-day band') are supported by evidence from extensive and systematic documentation of orientations at sites throughout Central Mexico. Sites were located and important structures within those sites were oriented from as early as the early 3rd century BCE in accordance with a horizon reference system defined by the dates of the zenith and nadir passages and related calendrically significant solar events of that 260-day band. Subterranean chambers and other constructed devices have been identified that enabled precise determination of the actual dates of those zenith passages and thereby permitted calibration of the horizon reference system from one solar cycle to the next. In these respects, the zenith sun of the 260-day band served as an organizing principle of the constructed space and calendrics of Central Mexico.

Introduction
In 1577 CE, Sahagun wrote of the Mesoamerican 260-day divinatory count that 'it is neither based on the influence of the stars nor on any natural thing'.[1] This view has changed little over the past 400+ years.[2] While some scholars have proposed that the 260-day *tonalpohualli* may have some relationship to the 260-day interval between solar zenith passages that

---

[1] Fray Bernadino de Sahagun, *Historia general de las cosas de Nueva España,* World Digital Library (Laurentian Library), available at
http://www.wdl.org/en/item/10615/zoom/#institution=laurentian-library&group=1&page=2&zoom=1.2856&centerX=0.6396&centerY=0.5250,
Prologue to Book IV; translation by author [accessed 10 November 2014].
[2] See, e.g., Mary Miller and Karl Taube, *The Gods and Symbols of Ancient Mexico and the Maya* (London: Thames & Hudson, 1993), p. 48; John S. Henderson, *The World of the Ancient Maya*, 2nd ed., (Ithaca: Cornell University Press, 1997), p. 50; David Stuart, *The Order of Days* (New York: Harmony Books, 2011), p. 154.

Harold H. Green, 'Zenith Sun as Organizing Principle of the Constructed Sacred Space and Calendrics of Central Mexico', *The Marriage of Heaven and Earth*, a special issue of *Culture and Cosmos*, Vol. 20, nos. 1 and 2, 2016, pp. 59–78.
www.CultureAndCosmos.org

occurs uniquely within the narrow latitudinal band centred on 14.72°N latitude (hereafter, '260-day latitudinal band' or simply '260-day band'; see Fig. 1), this proposition has been regarded as speculative because it has not been supported by convincing evidence and has therefore not gained acceptance.[3] Additionally, it has been argued that no satisfactory explanation of how the tracking of time could have been accomplished over a prolonged period without intercalation of days in the calendrical system which, as is generally agreed, did not occur.[4]

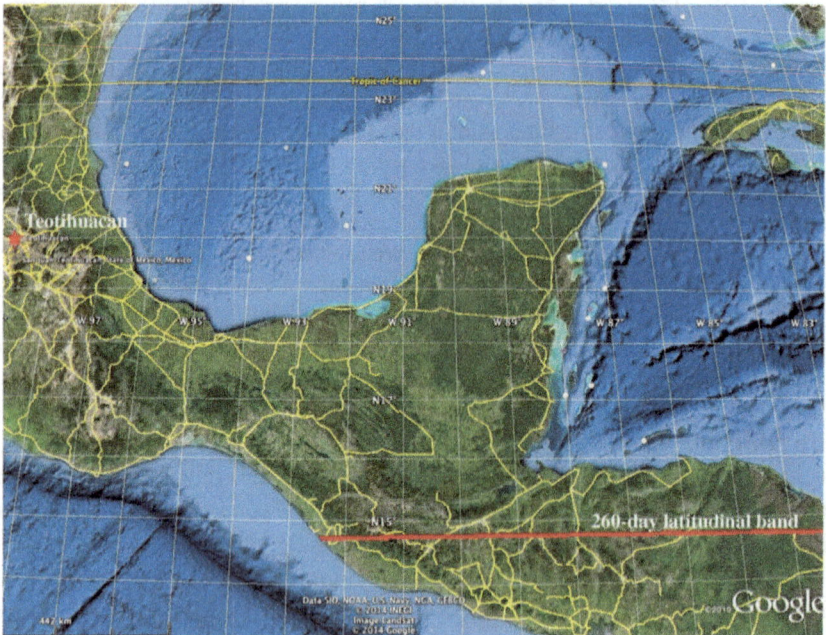

Fig. 1. Mesoamerica, showing the location of the 260-day band in relation to Teotihuacan. Google Earth image.

---

[3] See, e.g., Vincent Malmström, 'A Reconstruction of the Chronology of Mesoamerican Calendrical Systems', *Journal for the History of Astronomy* 9 (1978): pp. 105–16, 197; Johanna Broda, 'Calendrics and Ritual Landscape at Teotihuacan: Themes of Continuity in Mesoamerican Cosmovision', in *Mesoamerica's Classic Heritage*, ed. David Carrasco, Lindsay Jones, and Scott Sessions (Niwot, CO: University Press of Colorado, 2000), pp. 397–432.

[4] Ivan Šprajc, *Orientaciones astronómicas en la arquitectura prehispánica del centro de México* (Mexico, DF: Instituto Nacional de Antropologia e Historia [Coleccion Cientifica 427], 2001), pp. 135–46.

Broda, citing '[s]ystematic studies of alignments' in Mesoamerican architecture, has suggested avenues of research to answer these enduring but unresolved questions. She points specifically to the 'relevance of the latitude of 15°N' because of its 'conspicuous calendrical periods' and 'calendrical properties', and proposes that awareness of these calendrical properties had spread throughout Mesoamerica at least by the Classic Period.[5] She argues that 'in Mesoamerican cosmovision the greatest importance was attributed to the dates themselves, i.e., the precise days of the solar cycle', and that these 'most important dates' of the sun's annual course were tracked on the local horizon by means of a 'horizon reference system'.[6]

Ancient inhabitants of Central Mexico used the sky – more particularly, the sun marked at the local horizon – to locate and orient their sacred space in a manner significantly related to their means of tracking time, as established by orientation data systematically documented by Šprajc at 37 archaeological sites throughout Central Mexico.[7] This corpus of orientation data – comprised of both architectural orientations and horizon orientations – provides direct evidentiary support for a convincing argument that the 260-day *tonalpohualli* is directly related to the 260-day zenith passage interval, and that the Central Mexico 'horizon reference system' proposed by Broda was defined, as early as the third century BCE, by a set of fixed tropical year (Gregorian) dates that match, within reported margins of error, the dates of the zenith passages, the 'mirror opposite' nadir passages and related calendrically significant events that are unique to the 260-day band.[8] As will be developed more fully below, these calendrically

[5] Johanna Broda, 'Zenith Observations and the Conceptualization of Geographical Latitude in Ancient Mesoamerica: A Historical Interdisciplinary Approach', in *Viewing the Sky Through Present and Past Cultures*, edited by Todd W. Bostwick and Bryan Bates, (Phoenix, AZ: City of Phoenix, 2006), pp. 183, 188, 198.

[6] Broda, 'Calendrics and Ritual Landscape', pp. 398, 408. Since 'the dates of the precise days of the solar cycle' that she refers to are 'fixed by the alignments', the reference is necessarily to Gregorian dates; dates of the 360+5 day count (*xiupohualli*) are not fixed in the tropical year, and corresponding dates in the 260-day count (*tonalpohualli)* would vary significantly from one solar cycle to the next.

[7] The 37 sites include 8 Preclassic sites, 9 Classic sites, and 20 Postclassic sites. Selected sites were limited to those where the preservation of architectural remains permitted measurement of orientations with sufficient precision (Šprajc, *Orientaciones*, p. 55).

[8] Unlike solar zenith passage, occurring directly above the observer's head at noon, the 'mirror opposite' nadir passage, occurring under the observer's feet at midnight, is not directly observable. However, nadir sunset occurs 180° from zenith sunrise at a flat horizon, as can be observed from the top of a volcano – see

significant dates include the chronological midpoints between zenith
passage and equinox (10 April, 2 September), and between nadir passage
and equinox (1 March, 12 October).[9] They also include dates that are
separated by intervals of 13 or 26 days from the zenith passages, the nadir
passages, the zenith/equinox midpoints, or the nadir/equinox midpoints
(the four 'reference poles').

The actual dates of the 260-day band zenith passages were determinable
by means of subterranean chambers and pyramid shafts in Central Mexico,
designed and constructed so as to be capable of precise measurement of the
interval between the 260-day band zenith passages – sufficiently precise,
indeed, to detect the quadrennial increase of that interval to 261 days.[10]
Knowing the dates of the 260-day band zenith passages, a practitioner
familiar with the mechanics of the *tonalpohualli* would be able readily to
determine all other dates in the horizon reference system.

These orientation and observatory data establish that the unique
calendrical properties of the 260-day band had become shared knowledge
in Central Mexico, at least by the Classic period as proposed by Broda; that
the location of sites must have been determined by the spot where intended
alignments marking dates of 260-day band events could be achieved; and
that zenith sun of the 260-day band served as an organizing principle of
both the constructed sacred space and calendrics of Central Mexico.

**Fundamental differences between tropical and temperate zone
astronomy**
It is important preliminarily to emphasize certain fundamental differences
between temperate zone astronomy and tropical astronomy.

> [N]early all tropical cultures that developed indigenous astronomical systems…
> gravitated toward a reference system consisting of zenith and nadir as poles and
> the horizon as a fundamental reference circle. Such an arrangement stands in

---

Karen Bassie-Sweet, *Maya Sacred Geography and the Creator Deities* (Norman,
OK: University of Oklahoma Press, 2008), pp. 254–55.
[9] These marked dates are calendrically significant because they are separated by
intervals of 20 or 13 days (or multiples thereof) that are the factors of the
tonalpohualli.
[10] Broda, 'Calendrics and Ritual Landscape', p. 415.

remarkable contrast to the celestial pole-equator (or ecliptic) systems developed by ancient civilizations of the temperate zone.[11]

Zenith and nadir passages of the sun occur only within the tropics. The dates on which these solar events occur vary with latitude, and the dates on which they occur within the 260-day band are unique to that band.

Unique calendrical properties of the 260-day latitudinal band

In Mesoamerica, only within the 260-day band is the long interval between solar zenith passages 260 days, comprised of the factors 20 and 13 so intimately connected with Mesoamerican timekeeping. The dates of zenith passage are 30 April and 13 August, the latter date being the same as the base date of the Maya Long Count, 13 August 3114 BCE, also a zenith passage day only within the 260-day band.[12]

Only within that band is the time between the zenith and nadir passages equal to 80 days (4x20), and only there do the equinoxes divide this 80-day period into two 40-day periods.[13] The chronological midpoints between zenith passage and equinox and between equinox and nadir passage divide each of the 40-day periods into two 20-day periods, numerically equal to a fundamental period of Mesoamerican calendrics.[14] These 'mirror opposite' events are not redundant, but are an expression of the fundamental Mesoamerican concept of duality.

> One of the basic structural principles of Mesoamerican religious thought is the use of paired oppositions. In these pairings, there is a recognition of the essential interdependence of opposites... [C]ommon oppositional pairings include life and

---

[11] Anthony F. Aveni, 'Tropical Archaeostronomy', *Science* 213, no. 4504 (1981): pp. 161, 167, 168.
[12] According to the widely accepted 584,285 GMT correlation constant.
[13] Broda, 'Zenith Observations', pp. 187–88; Harold H. Green, 'Cosmic Order at Chocola: Implications of Solar Observations of the Eastern Horizon at Chocola, Suchitepequez, Guatemala', in *Archaeoastronomy and the Maya*, edited by Gerardo Aldana y Villalobos and Edwin L. Barnhart (Oxford: Oxbow Books, 2014), pp. 24, 27 (Table 1.1), p. 29 (Figures 1.10 and 1.11).
[14] The 'zenith/equinox midpoint' is marked by the stelae 12–10 baseline at the Classic Maya site of Copan, located within the 260-day band – see Anthony F. Aveni, *Skywatchers* (Austin: University of Texas Press, 2001), p. 254 – and the 'nadir/equinox midpoint' is marked by Volcan Atitlan at Preclassic Chocola, see Green, 'Cosmic Order', pp. 24, 27 (Table 1.1), p. 29 (Figures 1.10 and 1.11).

death, sky and earth, zenith and nadir, day and night, sun and moon, fire and water.[15]

The significance of the zenith/nadir duality in relation to Mesoamerican cosmovision is embodied in what Farriss calls the Mesoamericans' 'preoccupation with order':

> The key, I think, to the Mesoamericans' conception of time and to their entire cosmology is their preoccupation with order and above all with cosmic order. ... Time is part of the cosmic order. Indeed, for the Maya and the rest of Mesoamerica, time is cosmic order... [16]

The horizon reference system exemplifies this preoccupation with cosmic order, and the interrelation of the 260-day zenith passage interval and the 260-day calendrical count demonstrates how 'time is part of the cosmic order'.

Calendrical properties unique to the 260-day band are summarized in Figure 2.[17]

Central Mexico orientation data

The orientations documented by Šprajc at 37 archaeological sites throughout Central Mexico provide abundant evidence not only of a consciousness of the unique calendrical properties of the 260-day band, but of a mastery of that knowledge employed to define a 'horizon reference system' based upon those properties. Šprajc measured both architectural orientations – orientations of groups or structures to sunrises/sunsets on the horizons – and horizon orientations, orientations from groups or structures to sunrises/sunsets behind prominent peaks on the horizons.

---

[15] Miller and Taube, *The Gods and Symbols*, p. 81; See also Anthony F. Aveni, 'Maya Numerology', *Cambridge Archaeological Journal* 21, no. 2 (2011): pp. 187–216, on the extensive use of 'mirror symmetries' in the intervallic patterning of the Maya codices.

[16] Nancy M. Farriss, 'Remembering the Future, Anticipating the Past: History, Time, and Cosmology among the Maya of Yucatan', *Comparative Studies in Society and History* 29, no. 3 (July 1987): p. 574.

[17] Because of the system of intercalating leap days to maintain the Gregorian calendar in concordance with the tropical year, these dates can vary by ±1 day (Šprajc, *Orientaciones*, p. 159, footnote 1).

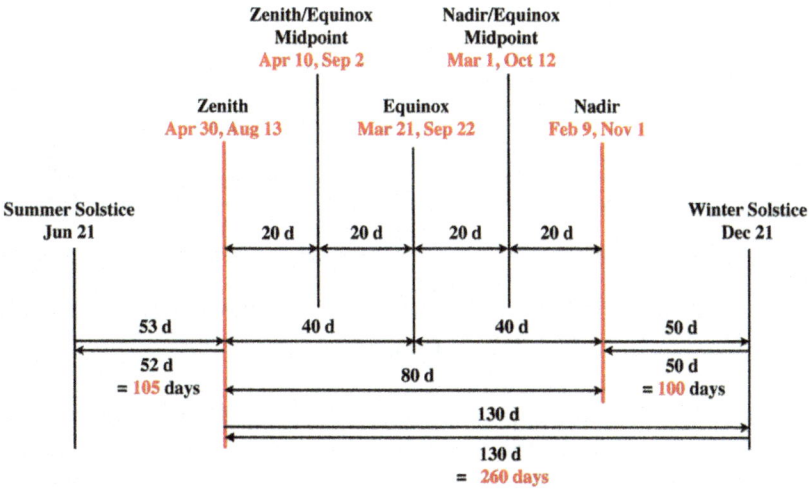

Fig. 2. Unique calendrical properties of the 260-day latitudinal band.

Interpretation of the orientation data

The reported architectural orientations of civic-ceremonial structures at each site are presented in tables that identify the specific groups or structures; the mean values of the azimuths (A) of the east-west axes measured along different lines physically preserved in the group or structure; the angular heights of the horizon (h) corresponding to the reported azimuths; the solar declinations (δ); and finally the dates corresponding to the declinations.[18] For all architectural orientations, Šprajc assigns to each of these elements an estimated margin of error. For example, the architectural orientations reported for the Pyramid of the Sun at Teotihuacan (see Fig. 3) are shown in the first four columns of Table 1.[19]

The margin of error for the mean azimuths is ±1°, equivalent to ±3 days for the dates marked at the east and west horizons. Where dates of a calendrically significant solar event within the 260-day band are within the margin of error estimated for the reported dates, those dates will be assumed to be what was originally intended to be marked by the orientation. In the above example, the dates of the 260-day band nadir passages (9 February and 1 November) are within the reported margins of

---

[18] Šprajc, *Orientaciones,* pp. 158–160.

[19] Šprajc, *Orientaciones*, p. 204, Table 5.34. All dates identified in this paper that exactly match dates of calendrically significant 260-day band solar events are represented in **bold** type.

error for the sunrise dates marked by the east-west orientation of the
Pyramid of the Sun.

Azimuth, A	Height, h	Declination, δ	Reported dates	Dates within margin of error	Solar event within 260-day band
105°45'±1°	2°01'±5'	-14°11'±1°	10 Feb, 30 Oct ±3d	9 Feb, 1 Nov	Nadir Passage
285°45'±1°	0°22'±10'	14°48'±1°	**30 Apr, 13 Aug ±3d**	**30 Apr, 13 Aug**	**Zenith Passage**

Table 1.

For horizon orientations, Šprajc measured the azimuth and angular height
from the principal structures at each site, preferentially temples, to each
prominent peak on the east and west horizons within the angle of the sun's
annual displacement.[20]

Sunset on
Apr 30, Aug 13   ←   285°45'±1°

Fig. 3. Left, The Pyramid of the Sun at Teotihuacan, Mexico, as viewed from the
Pyramid of the Moon (photo by author); right, a portion of the street grid at
Teotihuacan, after René Millon, *The Teotihuacan Map*.[21]

---

[20] Šprajc, *Orientaciones*, pp. 35, 47. As most of these orientations were measured
with theodolite, with an estimated error of 5', no corresponding date error is
stated.
[21] René Millon, *The Teotihuacan Map* (Austin: University of Texas Press, 1973),
Vol. 1, Part 2, Map 1.

Šprajc additionally recognizes, independent of the margin of error estimated for each measured architectural orientation, that:

> a tolerance of ±1 day should be considered for all dates since, due to the system of intercalations employed in the Gregorian calendar to maintain it in concordance with the tropical year, the same event – the sunrise/sunset along the same alignment – does not necessarily occur each year on the same date.[22]

This additional adjustment will be applied to the reported data. In the above Pyramid of the Sun example above (Table 1), if the estimated margin of error for the sunrise orientation had been ±1 day instead of ±3 days, the sunrise dates would exactly match those of the 260-day nadir passages with application of the ±1 day tolerance:

Reported dates	Dates within margin of error	Dates within ±1 day tolerance	Solar event within 260-day band
10 Feb, 30 Oct ±1d	9 Feb, 31 Oct	9 Feb, 1 Nov	Nadir passage
30 Apr, 13 Aug ±1d	30 Apr, 13 Aug	30 Apr, 13 Aug	Zenith passage

Table 2.

Considering only the margin of error of ±3 days, the 260-day band nadir passage dates and the reported dates of 10 February and 30 October are equally valid. However, the sunset orientation of the pyramid indisputably marks the dates of the 260-day band zenith passages, and an orientation to sunrise on 9 February and 1 November is the only way the orientation of the Pyramid of the Sun can embody the duality of zenith and nadir that Aveni has concluded is what defines the reference poles of the Mesoamerican astronomical system.[23] Further, as will be seen below, the dates most frequently marked by the Central Mexico orientation data are those of the 260-day band nadir passages (predominantly at sunrise) and zenith passages (predominantly at sunset).

When the marked dates reported by Šprajc are adjusted with respect to the reported margin of error and the ±1 day tolerance, a compelling pattern of dates emerges, as demonstrated in the following four sections.[24]

---

[22] Šprajc, *Orientaciones*, p. 159, footnote 1; translation by author.

[23] Aveni, 'Tropical Archaeoastronomy', pp. 161, 167, 168.

[24] The adjusted data corresponding to all orientations documented by Šprajc are presented comprehensively in tables that can be found in the Appendices to 'Zenith Sun as Organizing Principle' working paper by this author at the Academia.edu website,

Dates separated by intervals of 20-days

The reported dates marked by the architectural orientation of the Temple of
Ehecatl at the Postclassic site of Tecoaque, 1 May, 11 August ±2 days
(oriented to Cerro Chame) on the west horizon and 12 February, 30
October ±2 days (oriented to Cerro San Nicolas Sur) to the east, when
adjusted within the stated margins of error, are the dates of the 260-day
band zenith passages (Fig. 4, left) and nadir passages (Fig. 4, right),
respectively, demonstrating the same zenith/nadir duality as the Pyramid of
the Sun at Teotihuacan, and providing a good example of the symmetrical
'mirror opposite' event pattern that will be seen to emerge from the Central
Mexico orientation data.[25]

Fig. 4. Left, architectural orientation of the Temple of Ehecatl at Tecoaque to
sunsets behind Cerro Chame and horizon orientation from the Temple to sunsets
behind Cerro La Cantera; right, architectural orientation of the Temple to
sunrises behind Cerro San Nicolas Sur. Google Earth images.

---

https://www.academia.edu/19111816/Zenith_Sun_as_Organizing_Principle_of_the_
Constructed_Sacred_Space_and_Calendrics_of_Central_Mexico [accessed 2 July
2016].
[25] Šprajc, Orientaciones, p. 362, Table 5.144. The architectural orientation of a
'rectangular structure' at Tecoaque marks the dates (adjusted) of solar events that
occur ±26 days from the 260-day band nadir and zenith passages (3 February, 7
November and 6 May, 7 August, respectively) – see following section.

The horizon orientation to Cerro La Cantera on the west horizon marks the dates of the 260-day band zenith/equinox midpoints, 10 April and September 2 (Fig. 5, left).[26]

In the schematic of the Tecoaque orientations (Fig. 5), marked dates (in squares) are separated by intervals of 20 or 13 days (or a multiple). The dates marked by the architectural orientation of the rectangular structure on the east and west horizons (see footnote 25) and by the horizon orientations to Cerros El Rosario and San Nicolas Norte are symmetrically positioned on opposite sides of the zenith/equinox and nadir/equinox reference poles at calendrically significant intervals of 26 and 13 days, respectively.[27]

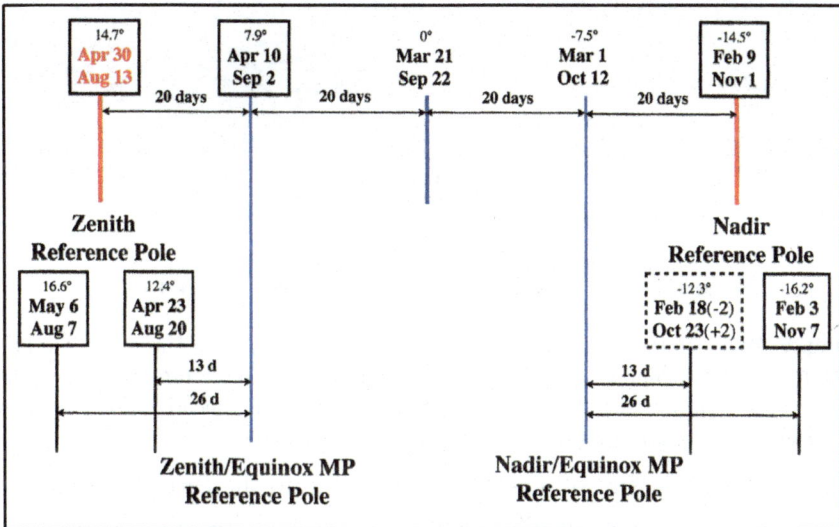

Fig. 5. Dates of 260-day band events marked by orientations at Tecoaque.

At the Preclassic site of Tx-TF-10, dated to the early third century BCE, the architectural orientation of the only surviving structure on Mound 119 marks the dates (when adjusted) of the 260-day band zenith and nadir passages, analogous to the orientations of the Pyramid of the Sun at

---

[26] Šprajc, *Orientaciones*, p. 363, Table 5.146.

[27] The dates 23 April and 20 August, ±13 days from the 260-day band zenith/equinox midpoints, are marked by the horizon orientation to Cerro El Rosario; dates for the San Nicolas Norte orientation are shown in a hatched square because they are ±2 days from the dates of the 260-day band event (16 February and 25 October).

Teotihuacan and the Temple of Ehecatl at Tecoaque.[28] The horizon
orientation to sunrises behind Cerro Tlamacas marks the dates of the
nadir/equinox midpoints, 1 March and 12 October (Fig. 6, right), a 'mirror
opposite' event to the horizon orientation to sunsets behind Cerro La
Cantera at Tecoaque previously mentioned (Fig. 6, left).[29]

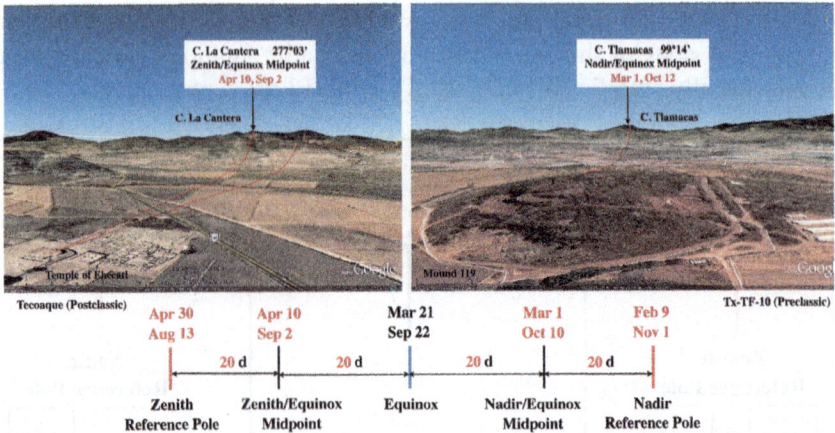

Fig. 6. Right, horizon orientation from Mound 119 at the Preclassic site Tx-TF-
10 to sunrises behind Cerro Tlamacas; left, 'mirror opposite' 260-day band event
marked by the horizon orientation from the Temple of Ehecatl at Tecoaque to
sunsets behind Cerro La Cantera. Google Earth image.

At the Preclassic site of Tx-TF-6, also dating to the early third century
BCE, architectural orientations mark the dates of the 'mirror opposite'
260-day band zenith and nadir passages. At the Classic site of Las Pilas,
Structures 1 through 4 are oriented to mark the dates of these same 'mirror
opposite' 260-day band events. Calendrically significant orientations at
Postclassic sites include the structures of Area F at Texcotzingo, Templo
Calendarico at Tlatelolco and Templo Mayor (stage II) at Tenochtitlan,

---

[28] Šprajc, Orientaciones, p. 193, Table 5.23. The horizon orientation to Cerro
Chicocuajio marks sunrises within ±1 day of 27 January and 14 November, ±13
days from the 260-day band nadir passages, and that to Pico Tres Padres marks
sunsets on 4 April, 8 September, ±26 days from the 260-day band zenith passages.
[29] Šprajc, *Orientaciones*, p. 193, Table 5.24.

oriented to mark the dates of the 'mirror opposite' 260-day band zenith/equinox and nadir/equinox midpoints.[30]

Fifty-eight (58) orientations mark dates of 260-day band events that are separated by intervals of 20 days (or multiple).

Dates separated by intervals of 13 and 26 days from the zenith and nadir reference poles

We have already seen at the Preclassic site of Tx-TF-10 horizon orientations marking (exactly or within ±1 day) dates of 260-day band events ±13 days from the nadir passages and ±26 days from the zenith passages (see footnote 28).

Examining the remaining Central Mexico orientation data will determine if orientations mark the date pairs that are 13 or 26 days from, and on opposite sides of, both the zenith and nadir 'reference poles'. Such 'mirror' symmetry is similar to the intervallic patterning in the Maya codices.[31]

Orientations from the Pyramid at the Postclassic site of Cerro de la Estrella, Mexico, D. F. mark sunrises behind Cerros Tlaloc, Telapon, Papayo and Guadalupe (Fig. 7).[32] The site was located, and the Pyramid constructed, so that peaks on the distant horizon would mark calendrically significant 260-day band dates on opposite sides of the zenith and nadir reference poles, at distances of 13 and 26 days.

Fifty one (51) orientations mark dates of 260-day band solar events that are 13 or 26 days from the zenith and nadir reference poles.

Dates separated by intervals of 13 and 26 days from zenith/equinox and nadir/equinox midpoint reference poles

On the west horizon at Teotenango, the orientation from Structures 1A to Volcan Nevado de Toluca marks sunsets on 28 March and 15 September

---

[30] See Šprajc, *Orientaciones*, p. 190, Table 5.18; p. 193, Table 5.23; p. 249, Table 5.49; p. 369, Table 5.151; p. 374, Table 5.156; p. 384, Table 5.161 (all reported dates adjusted within stated margins of error).

[31] Aveni, 'Maya Numerology', pp. 187–216.

[32] All are horizon orientations except for that to Cerro Guadalupe, which is the architectural orientation of the Pyramid. Another horizon orientation to Volcan Iztaccihuatl marks the dates 27 January and 14 November, ±13 days from the nadir reference pole. Horizon orientations to Cerros San Miguel and La Malinche mark sunset dates (within ±1 day) ±26 days from the nadir/equinox midpoint reference pole (dates adjusted for margins of error), and the dates of the local zenith passage (17 May and 26 July), respectively (Šprajc, *Orientaciones*, pp. 335–36, Tables 5.120, 5.121 and 5.122).

(Fig. 8, top).[33] At the site of Xochitécatl, Preclassic Structure E1 is oriented
to Volcan La Malinche, marking sunrises on 14 March and 29 September
(Fig. 8, bottom). These 'mirror opposite' orientations mark dates that are
on opposite sides of – and ±13 days from – the zenith/equinox and
nadir/equinox midpoint reference poles.

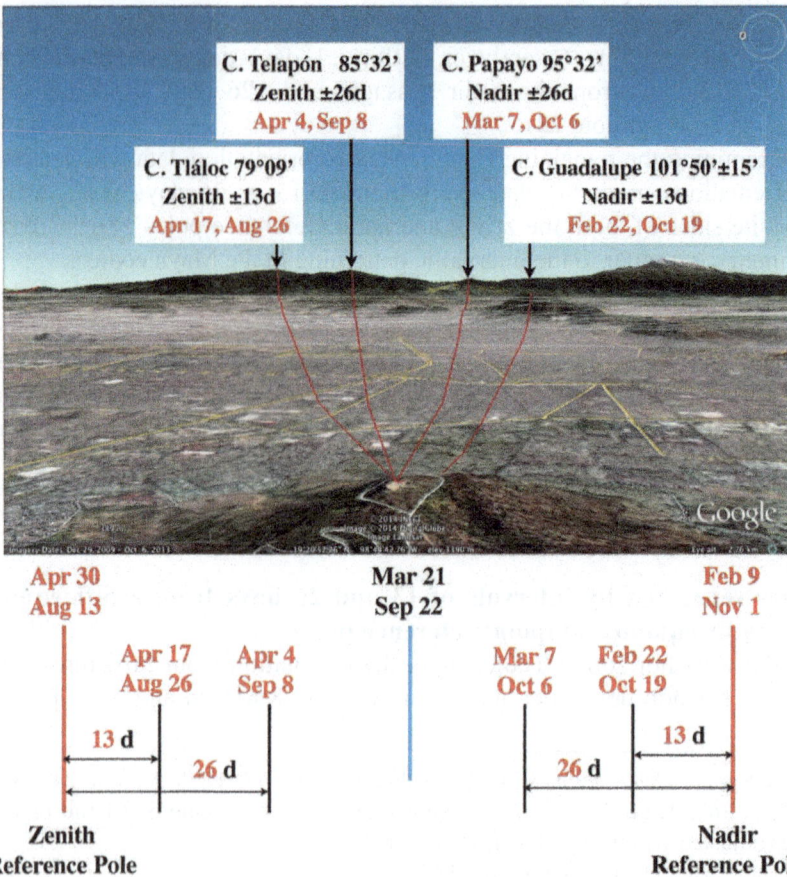

Fig. 7. The orientations from the Pyramid at Cerro de la Estrella to Cerros
Tlaloc, Telapon, Papayo and and Guadalupe. Google Earth image.

---

[33] A 'prominent peak' on the west horizon at Teotenango is likely outside the
angle of the sun's annual displacement, since its orientation was not reported by
Šprajc (*Orientaciones,* p. 289, Tables 5.80 and 5.82).

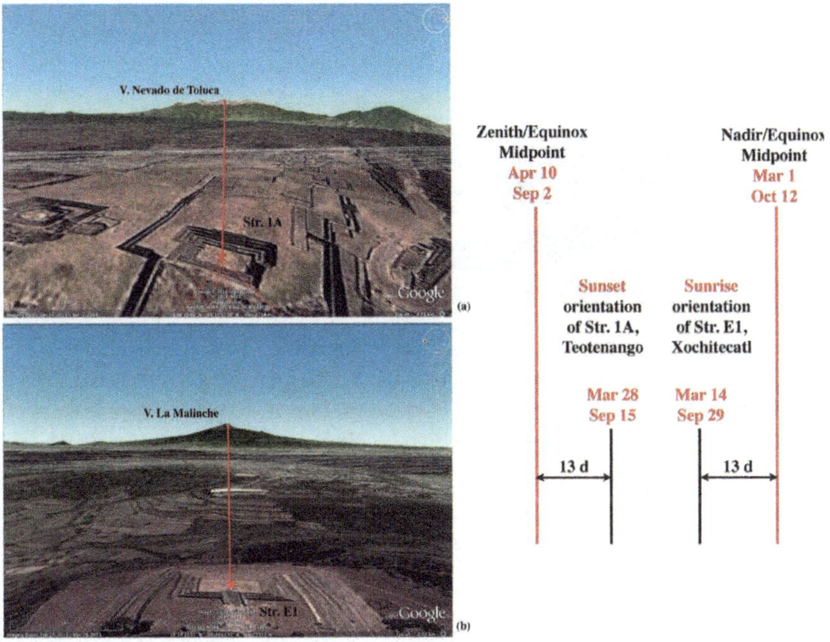

Fig. 8. Top, orientation from the Structure 1A at Teotenango (Postclassic) to sunsets behind Volcan Nevado de Toluca; and bottom, orientation from Structure E1 (Preclassic) at Xochitecatl to sunrises behind Volcan La Malinche. Google Earth images.

Thirty-five (35) orientations mark dates of 260-day band events that are 13 or 26 days from the zenith/equinox and nadir/equinox midpoint reference poles.

Other examples of calendrical symmetry
There are other examples of this positional/calendrical symmetry at Teotenango (Fig. 9).[34] Three orientations mark the dates 13 May and 31 July, except that one date is one day off (1 August instead of 31 July). Additionally, Structures 1A and 2D mark the same 'mirror opposite' events, except that the Structure 2D sunrise dates are also one day off (21 February and 20 October instead of 22 February and 19 October). Given the extent of the positional/calendrical symmetry that emerges from the Central Mexico orientation data, it is likely that all of these Teotenango orientations were intended to mark the dates of 260-day band events.

---

[34] Šprajc, *Orientaciones*, p. 286, Table 5.78; p. 289, Tables 5.79 and 5.81.

Structure or Horizon Marker	Declination, δ	Reported Dates	Dates within Margin of Error	Dates within ±1d Tolerance	Solar Event within 260-day Band
Str. 1B	18°20'±1°	May 12, Jul 31±4d	May 13, Jul 31	May 13, Jul 31	13d after NZP/before SZP
C. Mateo (1A&1B)	18°33'	May 14, Jul 31	May 14, Jul 31	May 13, Jul 31	13d after NZP/before SZP
C. Mateo (Str 2D)	18°03'	May 12, Aug 2	May 12, Aug 2	May 13, Aug 1(+1)	13d after NZP/before SZP(+1d)

Structure or Horizon Marker	Declination, δ	Reported Dates	Dates within Margin of Error	Dates within ±1d Tolerance	Solar Event within 260-day Band
Structure 1A	-10°15'±35'	Feb 22, Oct 19±2d	Feb 22, Oct 19	Feb 22, Oct 19	13d after NNP/before SNP
	12°02'±35'	Apr 21, Aug 21±2d	Apr 23, Aug 20	Apr 23, Aug 20	13d after NZP/E/before SZP/E MPs
Structure 2D	-11°09'±10'	Feb 19, Oct 22±1d	Feb 20, Oct 21	Feb 21(-1), Oct 20(+1)	13d after NNP/before SNP (±1d)
	12°56'±15'	Apr 24, Aug 19±1d	Apr 23, Aug 20	Apr 23, Aug 20	13d after NZP/E/before SZP/E MPs

Fig. 9. Examples of positional/calendrical symmetry of dates marked by orientations at Teotenango.

Forty-six (46) Central Mexico orientations correspond to one or the other of these two situations, and it is likely that, for each of these, the dates of 260-day band events were what the builders intended to mark.

Defining the horizon reference system of Central Mexico

Examining all of the Šprajc orientation data,[35] 82.2% (58 + 51 + 35 + 46 = 190/231) of the orientations – taking into consideration the reported margins of error for dates marked by architectural orientations, the ±1 day tolerance applicable to all dates, and the calendrical symmetries discussed in the previous section – mark calendrically significant 260-day band dates. The data thus convincingly demonstrate that 'the most important dates' of the annual course of the sun were the dates of the 260-day zenith and nadir passages and of the related calendrically significant solar events separated by intervals of 13 and 20 days (or multiple) from those

---

[35] Not including four non-solar orientations and orientations that mark the solstices and local zenith passages.

fundamental reference poles.[36] It is these dates that define the horizon reference system of Central Mexico (Fig. 10).

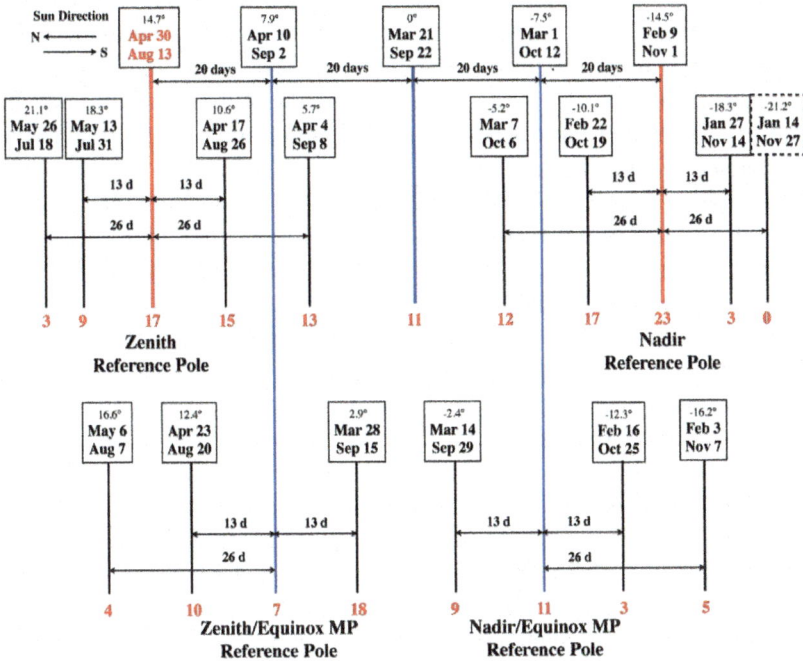

Fig. 10. The Central Mexico horizon reference system. The frequency with which date pairs are marked by orientations is indicated by numbers at the base of each reference pole; numbers above each date pair represent the approximate solar declinations.

All of the dates marked by the orientations are determined in relation to those of the 260-day band zenith passages: the dates of the other three reference poles and of the zenith/nadir midpoint (the equinoxes) are separated from the zenith passages by multiples of 20 days, and the dates marked by the other orientations are separated from one or the other of the four reference poles by 13 or 26 days. If the actual dates of the 260-day band dates were known, any practitioner familiar with the mechanics of the *tonalpohualli* would be able readily to determine all other dates in the horizon reference system.

---

[36] Broda, 'Calendrics and Ritual Landscape', p. 398.

Direct relationship between the 260-day zenith passage interval and the 260-day *tonalpohualli*

Zenith is the fundamental reference pole of indigenous astronomical systems in the tropics.[37] The Šprajc orientation data, as interpreted here, establish that it was the zenith of the 260-day band that defined the horizon reference system of Central Mexico which not only determined the location of sites and the orientation of their most important structures but also enabled the tracking of time.

Knowledge of the actual dates of the 260-day band zenith passages was essential in order to determine the dates marked by the horizon reference system from one solar cycle to the next. These dates were determinable by means of subterranean chambers (e.g., at Teotihuacan and Xochicalco) and vertical shafts (e.g., at El Tajin) constructed so as to be capable of precise measurement of the 260-day interval between the 260-day band zenith passages – sufficiently precise, indeed, to detect the quadrennial increase of that interval to 261 days.[38] Morante has determined, for example, that Cave 1 at Teotihuacan was constructed so that the first direct sunlight entered the chamber on 30 April and the last direct sunlight entered on 13 August, for a total of 105 days, with an interval of 260 days when the chamber was dark, that is, no direct sunlight entered; every four years that interval of relative darkness increased to 261 days.[39]

Counting in these chambers and shafts was by means of the 260-day *tonalpohualli*, and the *tonalpohualli* dates of the determined 260-day interval would be identical.[40] The *tonalpohualli* being used to determine zenith passage dates in these constructions, it follows that all dates of the horizon reference system were also *tonalpohualli* dates.

Thus, whatever may have been the origin of the 260-day count in archaic times, by at least the third century BCE there was a direct relationship between the 260-day zenith passage interval and the 260-day *tonalpohualli* embodied in the horizon reference system.

---

[37] Aveni, 'Tropical Archaeoastronomy', pp. 161, 167.

[38] Broda, 'Calendrics and Ritual Landscape', pp. 414–15;   Ruben B. Morante Lopez, *La Piramide de los Nichos de Tajin: Los codigos del tiempo* (Mexico, DF: Autonomous University of Mexico, 2011), pp. 130–33, Fig. 52.

[39] Broda, 'Calendrics and Ritual Landscape', pp. 414–15.

[40] Broda, 'Calendrics and Ritual Landscape', p. 415. Dates separated by an interval divisible by both 20 and 13 have identical day names and coefficients. Every fourth solar cycle, when the zenith passage interval would be determined to be 261 days, the date of the later zenith passage would be one day later in the *tonalpohualli* count.

Conclusions

The goal of this paper has been to demonstrate how the zenith sun of the 260-day band was used as an organizing principle for locating and orienting constructed sacred space and for tracking time in Central Mexico.

The most frequently marked dates at the horizons of Central Mexico sites are those of the 260-day band zenith and nadir passages: zenith passages are marked 14 times, predominantly at sunset; nadir passages are marked 20 times, predominantly at sunrise. The 260-day band zenith passages are marked by alignments far more frequently than local zenith passages, the latter being marked at only four sites (three times at Xochitecatl).[41] It was also clearly more important to Central Mexicans to mark the dates of the 260-day band zenith and nadir passages than those of the solstices, especially after the Preclassic; summer solstices are marked at only nine sites (five Preclassic) and winter solstices are marked at only seven sites (four Preclassic).[42]

The calendrically symmetrical pattern of marked dates that emerges from the Central Mexico orientation data provides direct archaeo-astronomical evidence that these sites were being carefully located and planned, both in relation to architectural alignment and orientation to prominent features on the horizon, to mark a specific set of dates at the local horizon by means of a region-wide horizon reference system defined by unique 'calendrical properties' of the 260-day band. These dates could not have been marked consistently by alignments throughout Central Mexico except on the basis of astronomical (solar) observations; otherwise, systematic divergences would have appeared at spatially or temporally separated sites. Indeed, the particular location of each of the site must have been determined by the spot where the intended alignments marking 260-day band dates could be achieved.[43]

---

[41] Šprajc, *Orientaciones*, pp. 183, Tables 5.10 and 5.11; p. 277, Table 5.67; p. 279, Table 5.70; p. 336, Table 5.122; and p. 345, Table 5.129. Local zenith passage could have been marked at other sites by other means, including by gnomon – see Vincent Malmström, *Cycles of the Sun, Mysteries of the Moon: The Calendar in Mesoamerican Civilization* (Austin: University of Texas Press, 1997), pp. 107, 135; Broda, 'Zenith Observations', pp. 198, 200.

[42] Šprajc, *Orientaciones*, ch. 5.

[43] See Johanna Broda, 'Astronomical Knowledge, Calendrics, and Sacred Geography in Ancient Mesoamerica', in *Astronomies and Cultures*, ed. Clive N. Ruggles and Nicholas J. Saunders (Niwot, CO: University Press of Colorado, 1993), pp. 280–81.

The actual dates of the 260-day band zenith passages, determinable in subterranean chambers and vertical shafts, enabled the dates marked by the horizon reference system to be adjusted from one solar cycle to the next. There was thus, at least by the third century BCE, a direct relationship between the 260-day *tonalpohualli* and the 260-day band zenith passages.

Sahagun can be forgiven for concluding in 1577 that the *tonalpohualli* was 'neither based on the influence of the stars nor on any natural thing'. Not until the extensive and systematic documentation by Šprajc of orientations at sites throughout Central Mexico, demonstrating remarkable consistency from the third century BCE through the Postclassic, has it been possible to define the horizon reference system in terms of the unique calendrical properties of the 260-day band, employed to locate sites and to orient important structures within those sites. And not until the identification of subterranean chambers and vertical shafts capable of precise determination of the actual *tonalpohualli* dates of the 260-day band zenith passages, essential for calibrating the horizon reference system from one solar cycle to the next, has it been possible to confirm convincingly the direct relationship between the 260-day zenith passage interval and the *tonalpohualli* of Central Mexico. The decipherment of this system of 'codes' that had been created in the 'living landscape' establishes that 'time and space were coordinated in the landscape by means of the orientation of buildings and settlements,' and that the zenith sun of the 260-day latitudinal band served as an organizing principle for both constructed sacred space and calendrics in Central Mexico.[44]

Acknowledgments
My grateful thanks to Nicholas Campion and the staff of the Sophia Centre for inviting me to present an abbreviated version of this paper at the 2014 Sophia Centre Conference, 'The Marriage of Heaven and Earth'; to Johanna Broda for giving me a copy of her 'Zenith Observations' paper that provided the inspiration to dig deeper into the importance of the zenith sun in Mesoamerica; to Ivan Šprajc for the extensive and systematically documented corpus of orientation data that made this study possible; and to Gerardo Aldana, Edwin L. Barnhart and Christopher Powell for their generosity in reviewing and making helpful comments on early versions of this paper. Responsibility for any and all errors and omissions is solely mine.

---

[44] Broda, 'Calendrics and Ritual Landscape', p. 398.

Communicating with the Ancestors in the Spiritual Landscape at Yaxchilán, Chiapas, Mexico

Stanisław Iwaniszewski

Abstract: For the Classic period Maya rulers, creating links with the past and the ancestors was important for various political and religious reasons. Maya rulers used different strategies to make visible connections to their ancestors, ranging from textual statements to iconographic representations. Very often they installed specialized buildings to create the complex web of different relationships existing between the city's patronal gods who controlled time and space, meteorological events and celestial bodies, and recent ancestors and current rulers. In a sense, ancestors 'configured' Maya cityscapes to mirror mythic narratives and the movements of celestial bodies in the night sky.

To recreate the dynamic relationship between materialized mythical places associated with the spiritual landscape, this paper explores astronomical alignments, visual connections and relative elevations between the ancient *otoot* (dwellings) structures at Yaxchilán, Chiapas, Mexico. As these structures served as loci for communication with the past, the present paper seeks to determine some of the organizational spatio-temporal principles determined by these royal *otoot*, such as astronomical alignments, designated times and calendar intervals that determined the cityscape at Yaxchilán. As some of the deified ancestors were transformed into heavenly bodies, these buildings became testimonials of royal ancestors reborn as celestial entities. In other words, the space defined by such buildings was a visible reminder of the presence of heavenly bodies perceived in night and day skies.

Introduction

Yaxchilán, the centre of the Classic Maya (ca. 350–810 CE) kingdom located on the left side of the Usumacinta River, the modern border between Mexico and Guatemala, is one of the Maya sites with very few structures oriented to the Sun. Though earlier investigations suggested that the city's main axes followed solar solstitial alignments,

Stanisław Iwaniszewski, 'Communicating with the Ancestors in the Spiritual Landscape at Yaxchilán, Chiapas, Mexico', *The Marriage of Heaven and Earth*, a special issue of *Culture and Cosmos*, Vol. 20, nos. 1 and 2, 2016, pp. 79–92.
www.CultureAndCosmos.org

archaeoastronomical research revealed scarce interest in solar alignments.[1] The lack of buildings with astronomically oriented alignments at Yaxchilán does not mean that events in the sky were not important to the daily life of its inhabitants. In this paper I argue that at Yaxchilán human-celestial relationships took a more nuanced, relational and complex shape permeated by shared cultural values and individual political actions.

The Yaxchilán's Late Classic rulers, Itzamnaaj Bahlam III (681–742 CE) and his son Bird Jaguar IV (752–768 CE), left a rich architectural and written legacy. In a public and grandiose fashion both rulers made major monumental renovations, rebuilt old structures, reset ancestral carved lintels, erected new buildings close to the ancestral ones, re-used the buildings attributed to the ancestors for commemorative ceremonies, and established variable ritual foci for connecting current rulers with both their recent ancestors and the mythical founders of their own dynasty.[2]

This paper explores the role of astronomical alignments in the context of the Late Classic concept of the royal *otoot* ('house', 'dwelling'). The Maya term *otoot* implied the house where someone lived, a kind of 'domicile' or 'dwelling', while another term *naah*, was used to refer to 'structure' or 'house' to describe such uninhabited structures like *ch'ulnaah*, 'sacred house, temple' (i.e., house made for the gods), *popol nah*, 'council building' etc.[3] *Otoot* structures associated with the rulers of Yaxchilán were built between 723 and 764 CE, encompassing the reigns of Itzamnaaj Bahlam III and Bird Jaguar IV.[4] Of five structures described as *otoot* dwellings (structures 10, 11, 22, 23, 44), two were associated with the ancestors (structures 22, 44) and maintained by actual rulers.

---

[1] Carolyn E. Tate, 'The Use of Astronomy in Political Statements at Yaxchilan', in *World Archaeoastronomy*, ed. Anthony F. Aveni, (Cambridge: Cambridge University Press, 1989), pp. 416–29; Carolyn E. Tate, *Yaxchilan: The Design of a Maya Ceremonial City* (Austin: University of Texas Press, 1992); Stanislaw Iwaniszewski and Jesús Galindo, 'La orientación de la Estructura 33 de Yaxchilán: una reevaluación', *Estudios de Cultura Maya* 28 (2006): pp. 15–26.

[2] Megan O'Neil, 'Object, Memory, and Materiality at Yaxchilan: The Reset Lintels of Structures 12 and 22', *Ancient Mesoamerica* 22, no. 1 (2011): pp. 245–69.

[3] David Stuart, '"The Fire Enters His House": Architecture and Ritual in Classic Maya Texts', in *Function and Meaning in Classic Maya Architecture*, ed. Stephen D. Houston (Washington, DC: Dumbarton Oaks, 1998), pp. 373–425; Shannon E. Plank, *Maya Dwellings in Hieroglyphs and Archaeology*, Vol. 1324, BAR International Series (Oxford: Archaeopress, 2004).

[4] Plank, *Maya Dwellings*, p. 35.

Today, the city is located at the apex of a horseshoe-shaped loop of the Usumacinta River. The Main Plaza, the city's ceremonial centre, was built on a narrow floodplain and runs parallel to the river from northwest to southeast, roughly aligned to the winter solstice sunrise. On the south and west other important structures were placed on the slopes and tops of the karst hills, composed of limestone formations, which rise over a hundred meters above the level of the river. This is a region of high rainfall, but with the dry season clearly marked between February and the first half May.

The ancestors in Yaxchilán

Creating material and symbolic links with the past and ancestors was extremely important for many Late Classic Maya rulers who, like Bird Jaguar IV at Yaxchilán, had problems of not being directly descending from the dynastic founders. As is known, an interregnum between the reigns of Itzamnaaj Bahlam III and Bird Jaguar IV lasted from 742 to 752 CE, though a short-lived ruler named Yopaat Bahlam II is mentioned, suggesting that he was the ruler who reigned at least during the part of this period.[5] Being a son of Lady Ik' Skull of Calakmul, his father's second and less important wife, Bird Jaguar IV had to prove his right to the throne. During his reign Bird Jaguar IV erected over 30 monuments and 3 Hieroglyphic Stairs, many of them referring to or displaying his ancestors.

Within the broader context of the ancient and modern Maya culture, the ability to communicate with deceased lineage founders and progenitors has been extremely important for their spirituality and religion.[6] Hieroglyphic texts that recorded events in the lives of particular rulers were re-used by the actual rulers to create and maintain the ties of descent, while ancestral representations in iconography served to legitimize rituals of royal succession.[7] Likewise, rulers often took names of their predecessors to revive an ancestor's vital essence in one of his descendants or to embody

---

[5] Simon Martin and Nikolai Grube, *Chronicle of the Maya Kings and Queens* (London: Thames & Hudson, 2008), pp. 127, 149; Martin and Grube, *Chronicle*, p. 127; and Flora Simmons Clancy, *The Monuments of Piedras Negras, an Ancient Maya City* (Albuquerque: University of New Mexico Press, 2009), pp. 158–59.

[6] Patricia A. McAnany, *Living with the Ancestors* (Austin: University of Texas Press, 2000), pp. 157–65.

[7] McAnany, *Living with the Ancestors*, pp. 39–40.

the calorific entity incorporated into the animated entity of royal lineages.[8]
During the Early Classic period the ancestors believed to be dynastic
founders were depicted as disembodied heads floating in the upper part of
the scene, a celestial abode. Late Classic ancestors were shown as full-
bodied individuals sitting above the scene, within clearly delineated upper
registers or in solar cartouches (e.g., in Palenque and Yaxchilán).

Bird Jaguar progenitors, Itzamnaaj Bahlam III and Lady Ik' Skull of
Calakmul, were portrayed as full-bodied ancestors sitting within solar and
lunar cartouches and referred to as being deified and solar and lunar in
nature (see Fig. 1)[9]. The placement of their images above the terrestrial
plane, framed by celestial cartouches, or portrayed in the guise of the lunar
and solar deities as the supervisors of the rituals performed by their
progenies, confirms that they were needed for political legitimization.[10]
Yet both progenitors were not depicted as being dead but in some way they
remained alive.[11] The gloss 'ub'aahil ['a'n], 'is representative of'', which
appears on Stela 11 (see Fig. 1), alludes to the animated entity called
b'aahis, 'head, forehead, face, body, being, image, portrait'.[12]

According to the Maya nobility concepts of self, a human being had at
least three animated entities roughly equivalent to the soul. One of them,
b'aahis, was immortal and did not travel to the Netherworld, but instead it
went to the heavens where resided with other Maya gods.[13] To conclude,
the figures of deceased rulers did not impersonate celestial deities (as the
rulers did during the public ceremonies performed in their lives), but rather

---

[8] O'Neil, 'Object, Memory, and Materiality', p. 249; Stephan Houston and David
Stuart, 'Of Gods, Glyphs and Kings: Divinity and Rulership Among the Classic
Maya', *Antiquity* 79, no. 268 (1996): pp. 289–312, here pp. 291–92.

[9] Tate, *Yaxchilan,* pp. 59–62; McAnany, *Living with the Ancestors*, p. 43.

[10] Tate, *Yaxchilan,* pp. 59–62; James L. Fitzsimmons, *Death and the Classic Maya
Kings* (Austin: University of Texas Press, 2009), pp. 120–23; and O´Neil, 'Object,
Memory, and Materiality', pp. 247–48.

[11] Harri Kettunen, 'Nasal Motifs in Maya Iconography' (Phd diss., Renvall
Institute, University of Helsinki, 2006), pp. 307–8.

[12] Stephan Houston and David Stuart, 'The Ancient Maya Self: Personhood and
Portraiture in the Classic Period', *RES Anthropology and Aesthetics* 33 (1998): pp.
73–101; and Erik Velásquez García, 'Los vasos de la entidad política de 'Ik': una
aproximación histórico-artística. Estudio sobre las entidades anímicas y el
lenguaje gestual y corporal en el arte maya clásico' (Phd diss., Facultad de
Filosofía y Letras, Universidad Nacional Autónoma de México, 2009), p. 561.

[13] Fitzsimmons, *Death and the Classic Maya Kings*, pp. 120–22; Velásquez
García, 'Los vasos de la entidad política de 'Ik'', pp. 560–61.

their animated entities were fused with the luminaries (men with the Sun, women with the Moon). In sum, at Late Classic Yaxchilán the sky becomes the dwelling of gods and of immortal entities of elite ancestors.

Fig. 1. The Bird Jaguar IV progenitors placed within solar cartouches: a) stela 4, b) stela 10, c) stela 1, d) stela 8, e) stela 11, f) stela 6. Drawings by Ian Graham (a, b, c and e) and Carolyn Tate (d and f).[14]

---

[14] Adopted from Ian Graham, *Corpus of Maya Hieroglyphic Inscription*, Vols. 2 and 3 (Cambridge, MA: Peabody Museum of Archaeology and Ethnology,

Houses and dwellings

It's probable that the Classic Maya at Yaxchilán believed a house was not merely a physical building but rather a living thing, possessing an animated entity similar to the human soul. When a new house was built or an old house was restored, the Maya performed rituals called 'fire entering' (*ochk'ahk' tuyotot*, 'the fire enters his house') consisting of intense incense burning within new buildings.[15] It is proposed that placing a fire inside a new building served to create its 'hearth', and heat itself reflected a house's vital forces or strength.[16] In other words, to bring the heat of fire into a house, meant that the building was being vivified and invested with its own animated entity. Briefly, through dedicatory ceremonies new or restored buildings acquired animated entities.[17] Even today, the modern Maya groups maintain that when the house is dedicated, it acquires a soul or *ch'ulel*.[18] In the light of this, actions like placing stelae and lintels with the names and images of past rulers inside *otoot* structures or portraying ancestors in stone sculpture attached to the buildings was a means of investing them with the animated entities of their owners.[19] Since *otoot* dwellings had their own names and those of their owners, and were related to the rulers who anchored them into the Long Count system, they should be interpreted as composite physical objects being constituted of parts, events and relationships.

Yaxchilán Emblem Glyphs

Yaxchilán rulers used two emblem glyphs. One form was read as *pa'chan* 'broken/split sky' (cleft design). Thus, the ruler's title would be read *k'uhul pa' chanajaw*, 'holy lord of split sky', while the name of the city

---

Harvard University, 1979 and 1982); and Tate, *Yaxchilan*, Fig 7, p. 36 and Fig. 71, p. 181.

[15] Stuart, '"The Fire Enters His House"', pp. 384–89.

[16] Stuart, '"The Fire Enters His House"', pp. 417–18.

[17] Brian Stross, 'Seven Ingredients in Mesoamerican Ensoulment: Dedication and Termination in Tenejapa', in *The Sowing and the Dawning: Termination, Dedication, and Transformation in the Archaeological Record of Mesoamerica*, edited by Shirley Boteler Mock, (Albuquerque: University of New Mexico Press, 1998), pp. 31–40.

[18] Evon Z. Vogt, *Zinacantan: A Maya Community in the Highlands of Chiapas* (Cambridge, MA: The Belknap Press of Harvard University Press, 1969), pp. 371, 461, 465; and Evon Z. Vogt, 'Zinacanteco Dedication and Termination Rituals', in *The Sowing and the Dawning*, Boteler Mock (ed.), pp. 21–30.

[19] See Houston and Stuart, 'The Ancient Maya Self', pp. 87–91, p. 95.

itself would be read *tahn ha' pa'chan*, 'in front of the water of split sky', 'mid-water split sky'.[20] It is still unknown whether this term refers to a particular mythic or local topography. Based on Tate's data, Canter proposed that the city's emblem glyph could have depicted the summer solstice sunrise through a notch between two peaks located in the Sierra de Lacandon some 20km northeast from Structure 41, one of the highest buildings at Yaxchilán, defining the visual line perpendicular to its façade.[21] However, measurements made *in situ* proved these alignments to be wrong.[22] Another emblem glyph was introduced by Itzamnaaj Bahlam III (681–742 CE): the royal title would be read as *k'uhul KAAJ ajaw*. 'holy lord of kaaj'.[23] Recent readings of this title suggest that *kaaj* refers to a mythical place associated with the Yaxchilán dynastic origins at the beginning of the present creation in 13.0.0.0.0 4 Ajaw 8 Kumk'u.[24] In one way or another, the installation of emblem glyphs reflects the claims of Yaxchilán rulers to establish a connection between current rulers and the origins of their dynasty.

Otoot buildings at Yaxchilán
So far, five structures at Yaxchilán have been described as *otoot* buildings[25]. Three structures were dedicated during the reign of Itzamnaaj Bahlam III (buildings 11, 23, and 44), while the other two were inaugurated by his son, Bird Jaguar IV (buildings 10 and 22). Table 1 summarizes the chronology of *otoot* structures in Yaxchilán.

Structure 44 occupies the central part of the northern edge of the West Acropolis. This building appears to commemorate military (i.e., captive-

---

[20] Simon Martin, 'A Broken Sky: The Ancient Name of Yaxchilan as Pa' Chan', *The PARI Journal* 5, no. 1 (2004): pp. 1–7, available at http://www.mesoweb.com/pari/publications/journal/501/BrokenSky.pdf [accessed 22 August 2016].

[21] Tate, 'The Use of Astronomy', p. 418, Fig. 32.2; Tate, *Yaxchilan*, p. 115; Ronald L. Canter, 'Rivers Among The Ruins: The Usumacinta', *The PARI Journal* 7, no. 3 (2007): pp. 1–24, available at http://www.mesoweb.com/pari/publications/journal/703/Usumacinta.html [accessed 22 August 2016].

[22] Iwaniszewski and Galindo, 'La orientación de la Estructura 33'.

[23] Alexandre Tokovinine, *Place and Identity in Classic Maya Narratives*, Studies in Pre-Columbian Art and Archaeology, n. 37 (Washington, DC: Dumbarton Oaks Research Library and Collection, 2013), p. 69.

[24] Alexandre Tokovinine, *Place and Identity*, pp. 73–74.

[25] Plank, *Maya Dwellings,* pp. 35–71.

taking) activities of the ruler, with texts that emphasize ancestral parallelism between actual military actions and the events of past Yaxchilán rulers. Structure 44 was dedicated on 7 Imix 14 Sots' date in 723 CE, possibly commemorating the accession date of the first Yaxchilán ruler, Yopaat Bahlam I on 8.16.2.9.1 7 Imix 14 Sots' (22 June 359 CE).[26] The number of days between these dates is 364 x 365 days.[27] The name of the building is 'Starry (or Four) Deer-Caiman'.

Structure	Dimensions (m)	Dwelling owner	Alignment 1	Alignment 2	Name
23	17.25 by 7.55	Lady K'ab'al Xook	32°01'	119°37'	In the middle of Pa'Chaan
44		Itzamnaaj Bahlam III	43°03'	127°32'	Starry Deer Caiman
11	17.3 x 4.5	Lady Sak B'iyaan	37°03'	130°09'	
10	17.45 x 5.50	Bird Jaguar	33°16' - 33°26'	123°18'	Heated Alligator?
22	18.5 x 2.7	Moon Skull	37°20'	126°36'	Four Bat Place

Table 1. Basic data of Yaxchilán *otoot* buildings. The site's average geographic latitude is 16°54' N, so the azimuth of solar sunrise events occurs between 65°26' and 114°34' (for an astronomical horizon). Structure dimensions according to Roberto García Moll, *La arquitectura de Yaxchilán.*[28]

Itzamnaaj Bahlam's wives were Lady K'ab'al Xook, the owner of Structure 23 which was dedicated in 726 CE, and Lady Sak B'iyaan, the owner of Structure 11, dedicated in 738 CE. Both structures are located at the Main Plaza and their dimensions are very similar (see Table 1). The name of Structure 11 remains unknown, while that of Structure 23 was Tan Ha' Pa' Chan, 'in front of the water/in the middle of Yaxchilán'; additional information from Lintel 26 describes Lady K'ab'al Xook as an impersonator of the Lunar Goddess.[29] The dedication of Structure 11 was made on 7 Imix date, recalling the 7 Imix dedication date of Structure 44 (the time between 7 Imix 14 Sotz and 7 Imix 19 Sip is 15 x 364 days). Both structures are described as the queens' 'houses' (*otoot*) but also as

---

[26] Mathews, *La escultura de Yaxchilán* (Colección Científica, 368) (México, D.F.: Instituto Nacional de Antropologia e Historia, 1997), p. 152; Martin and Grube, *Chronicle*, pp. 118–19.

[27] Plank, *Maya Dwellings*, p. 56.

[28] Roberto García Moll, *La arquitectura de Yaxchilán* (México, DF: Instituto Nacional de Antropología e Historia and Plaza y Valdés, 2003).

[29] Plank, *Maya Dwellings*, p. 40.

kabch'en of the ruler ('his lands and caves'), meaning that the ownership of both should be understood within the broader context of the ruler's authority over the lands, perhaps echoing the fact that both queens came to Yaxchilán from outside.[30] This interpretation may be supported by the fact that they did not dwell in buildings with strong links to the ancestors. As persons who were born in distant cities, they lacked direct familiar links to ancestral narratives and landforms.

Structure 23 is the burial place of Itzamnaaj Bahlam III (Tomb 2) and of his main spouse, Lady K'ab'al Xook (Tomb 3).[31]

Bird Jaguar dedicated two *otoot* structures (see Table 1). He built a new structure (Structure 22) in the place where the seventh Yaxchilán ruler called Moon Skull dedicated the building called the 'Four Bat House/Place'. To emphasize direct connections to the past this ruler gathered and mixed together old and new lintels.[32] As its dedication occurred 9 days after Bird Jaguar IV's own enthronement, it is probable that the ritual was part of his accession ceremonies.

On 9.16.13.0.0, 12 years after his accession, Bird Jaguar IV dedicated his own *otoot* (Structure 10) called *k'inichayiin?...peten*, 'heated alligator ?... island'.[33] The Mayan *k'inich* is derived from the adjective *k'ihnich*, often found in the names of the Maya rulers. It denotes heat as a vital quality ultimately derived from the Sun, *k'in*, but it is also accumulated with the advance of the age as well as with the offices of public or ritual functionaries.[34] All animate objects have their own *kìinil*. As Structure 10 approaches the Usumacinta River, the name may well refer to the nearby topographic features. The building was constructed upon a pre-existing platform that supported Structures 11 and 74.[35]

---

[30] Tokovinine, *Place and Identity*, p. 36–37.

[31] Roberto García Moll, Yaxchilán, Chiapas. *Arqueología Mexicana* 4, 22 (1996): pp. 36–45.

[32] Stuart, '"The Fire Enters His House"', p. 382; O'Neil, 'Object, Memory, and Materiality', p. 251.

[33] Plank, *Maya Dwellings*, p. 69.

[34] Velásquez García, 'Los vasos de la entidad política de 'Ik'', pp. 542–43; Søren Wichmann, 'The Names of Some Major Classic Maya Gods', in *Continuity and Change: Maya religious Practices in Temporal Perspective*, ed. D. Graña-Behrens, N. Grube, C. M. Prager, F. Sachse, S. Teufel, and E. Wagner (Acta Mesoamericana 14) (Markt Schwaben: Verlag Anton Saurwein, 2004), pp. 77–86.

[35] Plank, *Maya Dwellings*, p. 68.

Astronomical alignments of *otoot* buildings

At first glimpse, the spatial positioning of all *otoot* buildings at Yaxchilán is not astronomically functional (that is, all structures do not align to the positions of the Sun, Moon and planets on the horizon) (see Fig. 2 and Table 2). This is why I started to investigate how sunlight might be manipulated to illuminate the buildings' interiors. The ways in which sunlight worked as a significant component of Maya worldview might have been linked to its physical properties that produced dualisms like those between light and dark, and heat and cold. I used sunpath diagrams combined with the determination of the angle of sunray incidence throughout the day and over the seasons to predict sunlight penetration into the interior of buildings.

It is possible that the Classic Maya at Yaxchilán knew how the bright light of the Sun could have been manipulated and did not allow the sunlight to enter the interior in any great quantity. As is shown in Figure 2, in buildings that look toward the northeast (structures 22, 23, and 44), the sunlight penetrates the doors between April/May and July/August, while buildings facing southwest (structures 10 and 11), allow the sunlight to enter the interiors between September or October and February or April. In general, the northeast orientation of buildings allows the low-morning sun to penetrate deep into the interior, and the sunlight varies, being soft and diffuse at sunrise and sharp and direct when the sun disk goes sufficiently high to heat and dissipate the morning mist over the Usumacinta River. The southwest orientation of dwellings does not allow much direct sunlight inside the buildings. The western sky is often covered by rain-bringing clouds (especially between May and January), and the horizon is high, made up of a range of the hills located just a few hundred meters from the Man Plaza. All of this may greatly obstruct the view of the lowering afternoon sun, considerably shortening the period of sunlight duration.

There is an interesting match between Long Count dates referring to the dedicatory rituals and the moments of sunlight penetration into the buildings' interiors (see Table 2). Though the synchronization of Long Count dates with the proleptic Gregorian calendar cannot be regarded as being complete, the dates of buildings' dedication almost coincide with the periods of their interior illumination by sunrays.

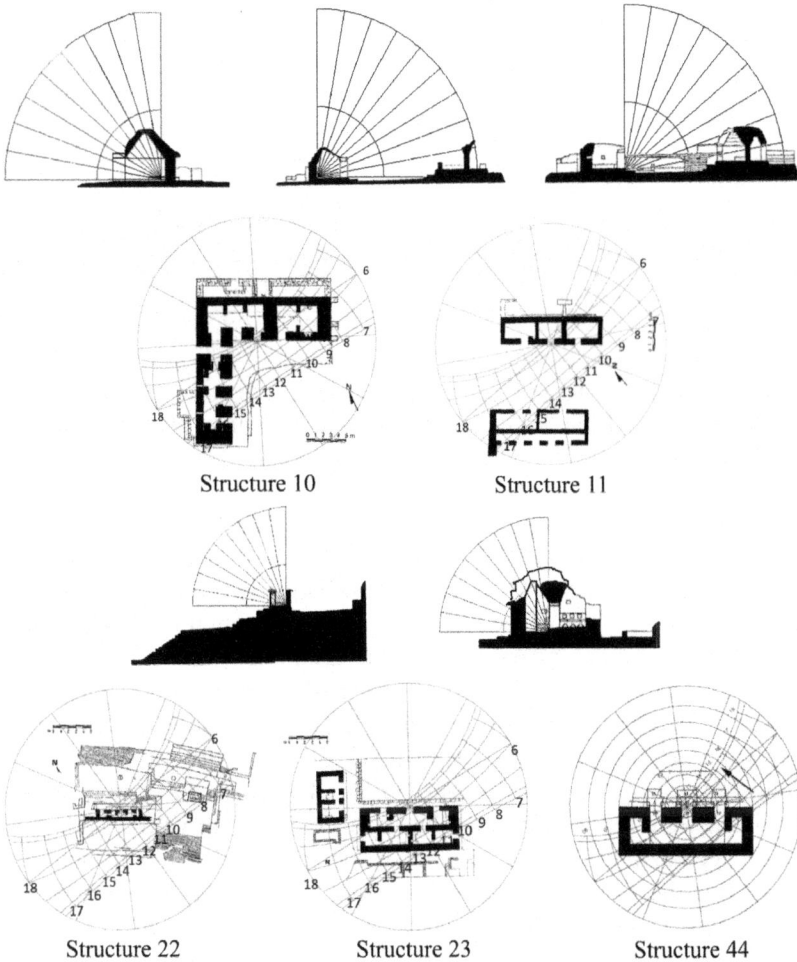

Structure 10   Structure 11

Structure 22   Structure 23   Structure 44

Fig. 2. Diagrams displaying sunlight illumination of *otoot* structures at Yaxchilán. Upper pictures show vertical sections through the buildings and the angles produced by the height of entrances enabling the interiors to receive direct sunlight; lower pictures show ground plans of these buildings with imposed sun path diagrams. There is no section image for Structure 44. Ground plans and vertical section of Structures 10, 11, 22and 23 redrawn from Moll, *La arquitectura de Yaxchilán*; ground plan of Structure 44 adapted from Ian Graham, *Corpus of Maya Hieroglyphic Inscriptions*.[36]

---

[36] Moll, *La arquitectura de Yaxchilán*; Graham, *Corpus of Maya Hieroglyphic Inscriptions*.

Structure	Monument	LC Date	Gregorian (584 283)	Event	Solar illumination
22	Lintel 21	9.0.19.2.4 2 K'an 2 Yax DD 9.16.1.0.9 7 Muluk 17 Tzek	16 Oct 454 12 May 752	*el –naah* house-censing	Early morning 21 Apr–22 Aug
23	Lintel 26	9.14.14.13.17 6 Kaban 15 Yaxk'in	26 Jun 726	*Ochk'ahk'* of Str. 23	Early morning 14 May–31 Jul
44	HS 3 Step IV	9.14.11.10.1 7 Imix 14 Zotz	27 Apr 723	*Och -k'ahk'*	Early morning 6 May–7 Aug
11	Lintel 56	9.15.6.13.1 7 Imix 19 Sip	8 Apr 738	*Och-k'ahk'* of Str. 11	Late afternoon 2 Sep– 10 Apr
10	Lintel 31	9.16.13.0.0 2 Ajaw 8 Wo	1 Mar 764	*Och-k'ahk'* of Str. 10	Late afternoon 14 Oct–28 Feb

Table 2. LC house dedicatory dates converted into the Gregorian and compared with the moments of interior illumination of *otoot* buildings at Yaxchilán. Solar illumination data are within ± 6 days (± 2°) of accuracy.

Of the five buildings Structure 44 seems to be the less connected to the Sun. The building is called 'Sky (or Four) Deer-Caiman' and is linked to the hieroglyphic stairway. One of its steps (Step III) bears the image of the Sun God, K'ihnich Ajaw (see Fig. 3), travelling through the interior of the Starry Deer-Caiman representing the underground or night sky.[37]

As a variant of the Cosmic Monster, it may represent the Milky Way, the path to the Otherworld.[38] It is noticed that the façade of this building is roughly perpendicular to the rising point of the Milky Way during the second half of April. On 23 April 723 CE, when the bright stars of Cassiopeia were rising between 4 and 5am, the contour of the Milky Way rose between azimuths of 29° and 42° respectively.[39] It was still before the sunrise and the Sun God was still carried on the back of 'Starry Deer Caiman'; this is why this building lacks strong solar alignments.

---

[37] Elizabeth Wagner, 'Mitos de creación y cosmografía de los mayas', in Nikolai Grube, ed., *Mayas, una civilización milenaria* (Köln: Könemann, 2001), pp. 281–93; David Stuart, *The Inscriptions from Temple XIX at Palenque* (San Francisco: The Pre-Columbian Art Research Institute, 2005); Eric Velásquez García, 'The Maya Flood Myth and the decapitation of the Cosmic Caiman', *The PARI Journal* 7, no. 1 (2006): pp. 1–10, available at (http://www.mesoweb.com/pari/publications/journal/701/Flood_e.pdf [accessed 22 August 2016].

[38] David Stuart, *The Inscriptions*, pp. 70–71.

[39] *Stellarium* ver. 0.12.4, data for the geographical location of Yaxchilán and the day of 23 April 723 CE.

Fig. 3. The image of K´ihnich Ajaw, the Sun God, within a cartouche placed on the body of Sky Deer-Caiman. Hieroglyphic Stairway 3 Step III. Drawing by Ian Graham. Adopted from Graham, *The Corpus.*[40]

Conclusions

At Yaxchilán alignment patterns did not reveal direct orientations towards solar and lunar movements along the horizon. However, due to the importance of the Sun to agriculture and elite metaphors concerning rulership, different strategies were used to mark rulers' connections with the sky. One way was to emphasize rulers' ties with his progenitors or ancestors, whose animated entities fused with the Sun and Moon gods. Rulers' personal identities were somewhat fused with other beings. Another way was to re-utilize the places/houses or calendar dates which were attributed to the activities of ancestors. These houses did not refer merely to the physical buildings, but rather to ancestral histories, place-names and calendar dates. In this way houses should be treated as material resources out of which take shape rulers' individual and elite corporate identities.

---

[40] Graham, *Corpus,* Vol. 3 Part 3, pp. 169–70.

The *otoot* iconographic symbolism related to the 'fire-entering' dedicatory ritual suggests still another way of study. Following Stuart's arguments on the meaning of fire-entering rituals, it may be concluded that bringing the heat of fire into a building was necessary to show that the place was vivified and invested with its own animated entity. [41] As noted earlier, modern Maya and Mesoamerican groups consider that heat (*k'in, k'ihnich, tonalli*) is related to person's animate entities. Curiously, *otoot* buildings located at the Main Plaza are physically heated by the rays of the Sun either early in the morning or late in the afternoon and those phenomena coincided with their dedicatory dates. I would argue that the animated entities of *otoot* dwellings, symbolized by their names and the presence of ancestor or actual owners, could have functioned as living beings because they received the calorific entity from K'ihnich Ajaw, or the Maya the Sun God.

Acknowledgments

This paper is derived from the project 'Starry sky – an animated sky' initiated during my sabbatical leave in 2014. My fieldwork was supported by CONACYT (Grant 25721-S 'Architecture and a spatio-temporal configuration in Maya society'). Thanks are due to the Consejo de Arqueología for respective permits.

---

[41] Stuart, '"The Fire Enters His House"', pp. 417–18.

The Joining of Heaven and Earth in Mormon Sacred Texts and Temples

Shon Hopkin

Abstract: As the only Christian faith tradition to currently build and use temples, Mormons view these buildings as modern 'mountains of the Lord', places where heaven meets earth. They also view the temple as a place that ritually centres and situates religious worshippers in their proper place in the cosmos. Accordingly, Mormon temples employ astronomical imagery to symbolically connect the temple as an earthly mountain with the order of the heavens. This imagery includes earth, moon, sun, and star stones on one of the first Mormon temples in Nauvoo, Illinois. It also includes moon and star stones on the iconic Mormon temple in Salt Lake City, Utah, with an image of the Big Dipper on one face of the temple, situated appropriately to point to the true north star in the skies above Salt Lake City. This paper will discuss Mormon astronomical temple imagery and its connection with literary themes in Mormon sacred texts that speak of the return of the city of Enoch and the joining of heaven and earth as a future destiny of the world, including statements that see this joining as a type of eternal marriage.

Members of The Church of Jesus Christ of Latter-day Saints, commonly known as Mormons, Latter-day Saints, or LDS, trace the organization of their church to 1830, with the publication of a book known as the Book of Mormon, provided by the church's founder and first prophet, Joseph Smith. This paper will trace the simple cosmology found in the Book of Mormon forward in time through increasing degrees of complexity in subsequent revelations provided by Joseph Smith. The only Christians today to actively worship in structures called temples, LDS understand these buildings as modern-day 'mountains of the Lord', a concept taken from Isaiah 2, where heaven and earth meet so that mankind and deity may engage in sacred encounters.[1] These temples employ astronomical imagery to describe the divine ascent of man back into the presence of God. Using primary sources, I will seek to demonstrate that the cosmology used in LDS temples weaves together various strands of cosmological thought

---

[1] See Joseph Blenkinsopp, Isaiah 1–39, in *The Anchor Yale Bible* (New Haven, CT: Yale University Press, 1974), p. 191.

Shon Hopkin, 'The Joining of Heaven and Earth in Mormon Sacred Texts and Temples', *The Marriage of Heaven and Earth*, a special issue of *Culture and Cosmos*, Vol. 20, nos. 1 and 2, 2016, pp. 93–112.
www.CultureAndCosmos.org

introduced by Joseph Smith throughout his lifetime, and that one of its goals is the joining of heaven and earth for the sake of LDS temple worshippers.[2] Through its use of imagery employing astronomical phenomena, modern LDS religious worship serves as a modern comparison with ancient faiths that incorporated heavenly imagery and orientation into their religious edifices and belief systems.

Fig. 1. San Diego, California LDS Temple.[3]

The earliest seeds of LDS cosmology are found in the faith's earliest sacred work, known as the Book of Mormon. Joseph Smith described the Book of Mormon as an inspired translation from records engraved by early inhabitants of the American continent, including God's dealings with them

---

[2] Although this study will focus on primary texts provided by Joseph Smith, two substantial studies on Latter-day Saint cosmology provide historical context for Smith's cosmological framework. See John L. Brooke, *The Refiner's Tale: The Making of Mormon Cosmology, 1644-1844* (Cambridge: Cambridge University Press, 1996); and Michael Quinn, *Early Mormonism and the Magic World View* (Salt Lake City, UT: Signature Books, 1998).
[3] All images by the author, excluding Fig. 2.

and a personal visitation to them by Jesus after his resurrection.[4] The book was intended to stand as a companion to the Bible, supporting and clarifying its teachings.[5] Like the Pentateuch (outside of the creation account in Genesis), little is found in its pages providing an overt cosmology. In two instances, however, the regular movements of heavenly bodies such as planets in relation to the earth are described as testifying plainly to the existence and power of a creator, based on a geocentric model. In one instance, a prophet named Alma is speaking with a dissenter: '...even the earth, and all things that are upon the face of it, yea, and its motion, yea, and also all the planets which move in their regular form do witness that there is a Supreme Creator' (Alma 30:44). Later, another Book of Mormon prophet uses an ancient, geocentric view of astronomical principles to show God's power:

> Yea, if he say unto the earth—Thou shalt go back, that it lengthen out the day for many hours—it is done; And thus, according to his word the earth goeth back, and it appeareth unto man that the sun standeth still; yea, and behold, this is so; for surely it is the earth that moveth [i.e. when God causes the stationary earth to move] and not the sun (Helaman 12:14–15).[6]

At the climax of the book, the voice of God is heard from the heavens, and Jesus subsequently descends to visit a large group of people who had assembled at a temple. Jesus heals, teaches, and ministers to the people

---

[4] Joseph Smith, *History of The Church of Jesus Christ of Latter-day Saints* [HC], 7 vols. (Salt Lake City, UT: Deseret Book Co., 1984), 1:315; or see Joseph Smith, 'History, 1838-1856, volume A-1 [23 December 1805-30 August 1834]' in *Joseph Smith Papers Project*, available at http://josephsmithpapers.org/paperSummary/history-1838-1856-volume-a-1-23-december-1805-30-august-1834?p=267&highlight=a%20record%20of%20the%20forefathers [accessed 12 November 2014].

[5] All references from the Book of Mormon, Pearl of Great Price (including the Books of Moses and Abraham), and Doctrine & Covenants are taken from the 2013 official edition of the Church of Jesus Christ of Latter-day Saints. Biblical quotations are also taken from the same edition, which uses the King James Version. All can be accessed at https://www.lds.org/scriptures?lang=eng.

[6] The geocentric reading of this text is described in David Grandy, 'Why Things Move: A New Look at Helaman 12:15', in *BYU Studies* 51, no. 2 (2012): pp. 99–128.

along with numerous angels that have also descended from heaven.[7] The people are so fundamentally changed that his visit ushers in a Zion-like community and era of peace lasting over two hundred years, mirroring the future millennial era of peace (as understood by Latter-day Saints) after Jesus's Second Coming.[8] The importance of a divine encounter centred on an ancient temple forms a foundation for later LDS temple and cosmological understandings. The Book of Mormon prophetic interest in the heavens, with astronomical principles being used to teach religious truths, would also act as a conceptual seed that would find fuller expression in later texts.

After the organization of the church in April 1830, Joseph Smith stated that he had been directed by God to work his way through the Bible, providing modern prophetic insight into biblical teachings that had been lost or altered over the centuries.[9] This work, known by LDS today as the Joseph Smith Translation, in some cases provided large additions to the biblical story, and some of these additions evidenced a greater interest in cosmology and the joining of heaven and earth. In June 1830, Smith provided what has come to be known as Moses chapter 1. The chapter acts as a preface to the abrupt beginning of Genesis's creation account, and as such shows interesting parallels to the Book of Jubilees, an account unavailable in English in Joseph Smith's day.[10] In the text, Moses is carried up into a high mountain upon which he has a theophany (an experience in which heaven and earth meet). He learns of the unspeakable grandeur of God and that he is a child of God created in God's image. After the vision, Moses is left to himself and is visited by Satan, who attempts to convince Moses that he is merely earthly, 'a son of man'.[11] Resisting this view and casting out Satan through God's name, another, grander heavenly vision opens before Moses's eyes.

---

[7] See 3 Nephi 11–18. Numerous prophetic figures take on the name of one Nephi, who was one of the first prophets in the Book of Mormon. Jesus' visit to the Book of Mormon occurs in the book (3 Nephi) named after one of these prophets.

[8] See Paul B. Pixton, 'Millennium', in *Encyclopedia of Mormonism* (New York: MacMillan, 1992), pp. 906–8, at
http://contentdm.lib.byu.edu/cdm/compoundobject/collection/EoM/id/4391/show/3893 [accessed 15 November 2014].

[9] Smith, HC, 2:18.

[10] James C. VanderKam, 'Jubilees, Book of', *Anchor Bible Dictionary,* 6 vols. (New York: Doubleday, 1992), 3:1030.

[11] Moses 1:12.

Moses is then shown the history of the world up until his time and the future of the world to the end of time. The vision is so extensive that, as the text states,

> As the voice was still speaking, Moses cast his eyes and beheld the earth, yea, even all of it; and there was not a particle of it which he did not behold, discerning it by the Spirit of God. And he beheld also the inhabitants thereof, and there was not a soul which he beheld not; and he discerned them by the Spirit of God.[12]

The vision does not stop there, but Moses goes on to see, 'many lands, and each land was called earth, and there were inhabitants on the face thereof'.[13] He is told that he is only to be concerned with the destiny of his own earth, but from 1830, LDS have believed as a scriptural tenet that there is life on other planets, and that such life is 'human', children of God.

When Moses asks why God has done all these things, God's response frames the heavenly vision within a comprehensible cosmology. In words that today are memorized by almost every Latter-day Saint, God tells Moses, 'Behold, this is my work and my glory, to bring to pass the eternal life of man'.[14] Thus the entire universe is understood as the place where God performs his work, and that work is centred on the salvation of his children, mankind. At the end of the chapter, Moses is told that the things that he has seen upon the mount are sacred, and are not to be shared openly with others, connecting his response to his vision with what would later become standard LDS temple practice, to keep the things revealed upon the 'mountain of the Lord' – that is, in LDS temples – sacred, not to be shared lightly or openly. This chapter then leads to the cosmogony of the Genesis creation account, which has now been framed as a portion of Moses's heavenly vision.

Joseph Smith continued to work through the book of Genesis, adding other portions of text at times to Adam's story, but most importantly to the story of Enoch, telling the story of Enoch's call as a prophet and Enoch's work to establish what is called the 'the city of Holiness, even Zion'.[15] Approximately seventy verses of text, according to modern divisions, detail a vision of Enoch, similar in broad strokes to Moses' vision, in which Enoch sees the history and the future of the world; he is shown the

---

[12] Moses 1:27–28.
[13] Moses 1:29.
[14] Moses 1:39.
[15] Moses 7:19.

power of Satan and is taught of the tender love of God for his creations. This love is ironically shown most poignantly at the moment when God weeps over the wickedness of his creations and chooses to flood the earth. A conversation between Enoch and God is recorded, in which Enoch learns that the earth is the most wicked of all God's creations throughout the cosmos,

> And Enoch said unto the Lord: ...Were it possible that man could number the particles of the earth, yea, millions of earths like this, it would not be a beginning to the number of thy creations;... how is it thou canst weep? The Lord said unto Enoch: Behold these thy brethren;...I can stretch forth mine hands and hold all the creations which I have made..., and among all the workmanship of mine hands there has not been so great wickedness as among thy brethren.[16]

Enoch's city is eventually translated, the entire population being taken up into heaven to dwell with God. Smith understood the word 'translated', found in the Hebrews 11:5 description of Enoch, as a state between mortality and glorified resurrection. Smith later taught that one of the roles of the translated beings from Enoch's city was to 'serve as ministering angels unto many planets'.[17] Since he also taught that this earth would receive no ministering angels from other planets (meaning that LDS do not believe in extraterrestrial visits to earth as a scriptural tenet), this concept added an additional layer to the earth's place in the cosmological framework: it is the most wicked of God's creations; some of its righteous inhabitants are translated and become ministering angels to other planets; but the earth does not receive any ministering angels from other worlds.[18] A final piece of this cosmological puzzle is apparently given in a vision Smith would have in 1832, recorded in Doctrine and Covenants 76, that will be discussed further below. In this heavenly vision, Smith stated that the inhabitants of all of God's worlds are saved by the same Jesus that saved this world, and by the same atonement worked out on this world.

---

[16] Moses 7:29–36.

[17] Smith, HC, 4:210; or see Smith, 'History, 1838–1856, volume C–1 [2 November 1838–31 July 1842]', *Joseph Smith Papers Project*, available at http://josephsmithpapers.org/paperSummary/history-1838-1856-volume-c-1-2-november-1838-31-july-1842?p=552&highlight=%22unto%20many%20planets%22 [accessed 16 November 2014].

[18] Doctrine & Covenants 130:5.

Although these conceptual dots were never explicitly connected by Smith, the cosmology introduced into his teachings demonstrates its own type of geocentricity. Jesus lived and died on this world, suffering here not only for the sins of this world's inhabitants but for all of God's creations through the entire universe, placing this planet at the centre of God's plan for his children. Inhabitants from this world would thus be appropriate as witnesses to other worlds, while this world would not necessarily need witnesses from elsewhere. While in one sense this cosmology is geocentric, it also further broadens the LDS view of the grandeur of God and of Jesus's atonement, providing a cosmological view that embraces all of the universe, rather than just this earth. In a time of increasing astronomical awareness, Smith worked to place the scope of Christianity within that broadened framework.

At the end of his vision, Enoch sees that the future of God's work upon this world would include the return of Enoch's own translated city. In the end of times, angels would descend to earth again to prepare a Zion-like people, led by a prophet driven by a longing to create Zion, and this latter-day people would meet Enoch's descending city in a true union of heaven and earth that would eventually witness the return of Jesus and the ushering in a millennium of heavenly peace. As the text states it, 'And righteousness will I send down out of heaven... to gather out mine elect from the four quarters of the earth... that my people may... be looking forth for the time of my coming; for there shall be my tabernacle, and it shall be called Zion...'.[19] Note should be taken here of the connection between the tabernacle or temple, the reunion of heaven and earth, and the creation of Zion (as in the Book of Mormon text).

> And the Lord said unto Enoch: Then shalt thou and all thy city meet them there, and we will receive them into our bosom, and they shall see us; and we will fall upon their necks, and they shall fall upon our necks, and we will kiss each other; And there shall be mine abode, and it shall be Zion...[20]

The return of Enoch was so important that, in the Joseph Smith Translation, God's provision of the rainbow to Noah was to act not only as

---

[19] Moses 7:62.
[20] Moses 7:63–65.

a sign that he would not flood the earth again, but more importantly, as a reminder that Zion would return.[21]

Latter-day Saints have often understood the earth as following the same pattern of righteousness as that commanded of God's disciples. The flood is the earth's baptism.[22] The fire of Jesus's second coming is the baptism by fire, or the gift of the Holy Ghost. The millennial era of peace is the earth's Sabbath of rest. The earth will die as a fallen world and be 'resurrected' as a celestial globe, as discussed in Revelation 21.[23] And Apostle Parley P. Pratt significantly described the joining of heaven and earth in this case as the earth's marriage.[24]

Before moving to a discussion of LDS temples and overtly-temple-connected texts, a vision of Joseph Smith with important cosmological implications, mentioned briefly above, must be discussed. In February 1832, a heavenly vision was opened to Smith and another LDS leader named Sidney Rigdon, in which they were shown that heaven was actually a realm composed of many varying degrees of glory, a description that fits with a loose interpretation of Paul's statement in 1 Corinthians 15:40–42. Those resurrected to a Celestial glory because of their valiance in accepting Christ's gospel would obtain bodies with a glory of the brightness of the sun.[25] Those who were honourable but not prepared to give all would still be saved to a glory termed 'Terrestrial', likened to the glory of the moon.[26] And those who accepted Christ in the end but did not show much in the way of true devotion would still be saved in a 'degree' or glory of salvation termed 'Telestial', with a glory likened to the stars as seen from earth.[27] Only those who completely rejected Christ would be damned to Outer Darkness (or Hell), with a complete lack of glory.[28] Thus

---

[21] Joseph Smith Translation [JST] Genesis 9:21-24. Located as footnote for Genesis 9:16a, at https://www.lds.org/scriptures/ot/gen/9?lang=eng [accessed 16 November 2014].

[22] Joseph Fielding Smith, *Doctrines of Salvation*, 3 vols. (Salt Lake City, UT: Deseret Book, 2014), 2:320.

[23] See Brigham Young, *Journal of Discourses*, 26 vols. (Salt Lake City, UT: Deseret Book, 1974), 1:155.

[24] Parley P. Pratt, *Voice of Warning* (1837; repr. Salt Lake City, UT: Deseret Book, 2001), pp. 95–97.

[25] Doctrine & Covenants 76:70.

[26] Doctrine & Covenants 76:71.

[27] Doctrine & Covenants, 76:81. Smith described the three broad divisions of salvation – Celestial, Terrestrial, and Telestial – as 'degrees' of glory.

[28] Doctrine & Covenants 76:30–38. See also Doctrine & Covenants 01:91; 103:73.

heavenly bodies were again brought to bear to help Latter-day Saints understand God's grand plan for mankind and their future destiny.

All of these cosmological concepts are brought to full fruition in a scriptural text Joseph Smith provided in 1842, that should probably be seen as a direct precursor to LDS temple ceremonies. Known as the Book of Abraham, the vision introduces another version of the creation account very similar to Genesis 1–3. This creation account is situated at the end of Abraham's quest to know God and receive power and authority at God's hands.[29] In the biblical account in Genesis, Abraham has a theophany in which he is told that his posterity would be greater than the stars of the heavens.[30] In the Book of Abraham, the patriarch sees the heavens and is led to an intimate understanding of the universe, and the world's place within that universe according to God's plans for mankind.[31]

This heavenly vision introduces what some have considered one of the strangest among many unique LDS scriptural teachings, the concept of a star named 'Kolob' that exists 'nigh unto the throne of God'.[32] The star is described at the centre of all things, with a time of revolution much slower than the earth upon which Abraham stands. It is the greatest of the 'governing ones' and gives forth light to all of God's creations throughout the universe.[33] A connected text states that Kolob is surrounded by fifteen other governing stars, and these heavenly bodies are at the centre of the orbit of all the stars in the heavens.[34]

The Kolob text is so matter-of-fact in its description, and the concept is so scandalously different than what Christians are accustomed to, that many Latter-day Saints often entirely miss the text's central point. Midway through the chapter, God likens his discussion of the great governing star Kolob to Christ's central role among God's children, the remainder of the stars.[35] The point of the astronomical concept of Kolob is to teach Abraham about his relationship with God, and – according to the text – to prepare him to teach the idolatrous Pharaoh religious truths using a

---

[29] Abraham 2:6–12.

[30] See Genesis 15:5.

[31] Abraham 3:1–17.

[32] Abraham 3:9.

[33] Abraham 3:3–5; 3:8–10.

[34] Abraham, Facsimile 2:5. See also Abraham 3:9. Brigham Young would later provide the cosmologically fascinating suggestion that the earth was created in the presence of Kolob, and only came to take up its current location after the fall of Adam. See Brigham Young, *Journal of Discourses*, 17:143.

[35] Abraham 3:18–19.

discussion of astronomy with which the Egyptians were so fascinated.[36]

Fig. 2. Book of Abraham, Facsimile 2. Hypocephalus.

God the Father sits on his throne, but Christ is Kolob, the great governing star closest to God that gives light to all of God's creations and whose 'slower' revolution time allows a Godly, eternal view. Other great leaders – the apostles, prophets, and others – orbit Christ and are set to govern and

---

[36] Abraham 3:15. That Abraham taught the Egyptians astronomical concepts is also found in the writings of Josephus, the Testament of Abraham, the Book of Jubilees, and the Qur'an. See Josephus, *Antiquities of the Jews*, tr. William Whiston, (Peabody, MA.: Hendrickson Publishers, 1987), 1.7.2; and Testament of Abraham 9–10, James H. Charlesworth, ed., *The Old Testament Pseudepigrapha*, 2 vols. (Garden City, NJ: Doubleday, 1983–85), 1:886–88; Jubilees 12:16–19, Jubilees 2:81; and Qur'an 6:76–80.

teach the rest of mankind. The number fifteen could be interpreted biblically as representing the three patriarchs and the twelve sons of Jacob, or in the context of the LDS church as the presiding first presidency and the quorum of twelve apostles. These governing ones hold places of importance but only as they themselves are found in orbit around Kolob.

To complete the lesson, Abraham sees in vision the pre-mortal spirits of mankind, called 'intelligences', some of whom are known as 'the noble and great ones'.[37] God stands 'in the midst' of these noble and great ones and says,

> 'We will go down, ...and we will make an earth whereon these may dwell; And we will prove them herewith, to see if they will do all things whatsoever the Lord their God shall command them; ...And they who keep their... estate shall have glory added upon their heads for ever and ever...'.[38]

Abraham then sees a council in heaven in which Lucifer (i.e., Satan) rebels against God and is cast down to heaven with his followers, while those who follow God accept Jesus as their Saviour pre-mortally and prepare to be sent to earth in their due time to fulfil their purpose.[39] This vision leads directly to a view of the creation of the world described in Abraham 4–5. With this understanding of the central role of Christ, the role of Satan, and an understanding of the purpose of this earth, all of it compared to the grand order of the heavens, Latter-day Saint readers are prepared to understand the story of the creation, Satan's presence in the garden, and the drama of the fall of Adam and Eve in very unique ways. Their own purpose and glory is again likened to heavenly bodies, as in the vision of the heavenly degrees of glory.

The Book of Abraham is Joseph Smith's translation of some papyri he purchased in 1835. Although most – but not all – of the papyri apparently burned in the great Chicago fire, the scriptural account Smith published had three images from the papyri copied in order to accompany the text.[40] All three were interpreted by him as illustrating the concepts taught in Abraham's story, although Egyptologists have shown that the three are actually connected to the ancient Egyptian religious texts known as the

---

[37] Abraham 3:22.

[38] Abraham 3:24–26.

[39] Abraham 3:27–28.

[40] H. Donl Peterson, 'Some Joseph Smith Papyri Rediscovered', *Studies in Scripture* Vol. 2, ed. Robert L. Millet and Kent P. Jackson (Salt Lake City, UT: Randall Book Co., 1985), pp. 183–93.

Book of Breathings and the Book of the Dead.[41] These Egyptian religious texts are designed to assist the deceased in his journey from death to resurrection, including final deification with the gods.[42] The second image, known to Egyptologists as a hypocephalus, is particularly interesting. Smith clearly designates it as a temple text,[43] and the Egyptian on it actually identifies it as connected to the Egyptian temple complex in Hieropolis.[44] Smith's interpretation places Kolob as the central image, with the rest of the images demonstrating the relationship of other heavenly bodies to Kolob. Interestingly, the Egyptian meaning of the hypocephalus was to demonstrate the daily journey of the sun through the realms of the living and the dead, symbolizing the divine journey of the deceased through life, into death, and into the presence of the gods.[45] Smith's description, seen in the light of the Kolob/Christ equivalent taught in the text, thus parallels the Egyptian use of the text in interesting ways. Smith's overt indication that the meanings of the image could only be understood in the temple also shows that he saw the temple ceremonies – which, as will be seen, centre on the divine ascent – as directly connected with this image.

The discussion finally arrives at LDS temples, which bring together the various strands of cosmological understanding developed earlier. When Joseph Smith designed the Nauvoo Temple, he provided moon stones, sun stones, and star stones. Similarly conceived stones were placed upon the Salt Lake City Temple by second church president Brigham Young.

Latter-day Saints who are not familiar with the concept of cosmology are often unaware of the earth and Saturn stones, and assume that the sun,

---

[41] See Michael D. Rhodes, 'A Translation and Commentary of the Joseph Smith Hypocephalus', *Brigham University Studies* 17 (Spring 1977): pp. 259–74, available at
https://docs.google.com/file/d/0B0LuDGvEmEgJM0g1dkxwRTlqME0/edit
[accessed 14 November 2014].
[42] See Hornung, *The Ancient Egyptian Books of the Afterlife*, trans. David Lorton, (Ithaca, NY: Cornell University Press, 1999).
[43] Book of Abraham, Facsimile 2, Note 8, states, 'Contains writings... to be had in the Holy Temple of God'.
[44] Michael D. Rhodes, 'The Joseph Smith Hypocephalus... Twenty Years Later', p. 5, available at
https://docs.google.com/file/d/0B0LuDGvEmEgJM0g1dkxwRTlqME0/edit
[accessed 14 November 2014].
[45] Alfred Wiedemann, *Religion of the Ancient Egyptians* (Dover, DE: Dover Publications, 2001), p. 306.

moon, and star stones are meant to demonstrate the three degrees of heavenly glory (as found in Doctrine and Covenants 76). They thus expect to find star stones at the bottom (for telestial glory), moon stones next (for terrestrial glory), and the sun stones at the top (for celestial glory), and are confused to instead see moon stones at the bottom (with earth stones below them on the Salt Lake Temple), then sun stones, and the star stones at the top. This order, of course, replicates the actual astronomical order in which the heavens are seen from the earth. The temple helps situate its worshippers in their real and proper location in the heavens, and helps lift them heavenward in their divine ascent toward God.[46]

Fig. 3. Nauvoo, Illinois LDS Temple. Restored according to building pattern in 1840's.

Other astronomical symbolism is also used to teach that the temple is designed to lead Latter-day Saints toward God. An image of the Big Dipper is found on the western face of the Salt Lake City Temple that actually points directly to the North Star in the sky, again situating the

---

[46] Paul Thomas Smith and Matthew B. Brown, *Symbols in Stone* (Salt Lake City, UT: Covenant Communications, Inc., 1997), pp. 149–56.

temple and its worshippers properly in place in the cosmos.[47] The North Star is not on the temple because the temple does not represent God; rather, the temple guides to God. The temple is not the centre. Rather, the temple points to God who is at the centre. This symbolic use of the stars connects closely with the centricity of the starry images in the Book of Abraham. The Nauvoo Temple provides a different star image with a similar purpose, a downward pointed star (as can be seen in Figure 7), that reminds many non-LDS observers of occult imagery.[48] According to LDS interpretation, the designed purpose of the downward star is to remind of the star that pointed the wisemen to Jesus's location, with the symbolic meaning, 'God can be found here'.[49] The Provo, Utah, LDS temple uses a different type of symbol to teach the same concept. It is designed to look like a cloud by day and a pillar of fire by night, the objects that led the children of Israel through the wilderness and at the centre of which were found Jehovah.

Fig. 4. Nauvoo Temple Moon Stones

[47] Smith and Brown, *Symbols in Stone*, pp. 156.

[48] See Val Brinkerhoff, *The Day Star: Reading Sacred Architecture, Book 2* (Salt Lake City, UT: Digital Legend Press, 2012), p. 391.

[49] Brinkerhoff, *The Day Star, Book 2*, p. 131.

Fig. 5. Nauvoo Temple Sun Stone.

Fig. 6. Nauvoo Temple Star Stones and Sun Stone.

Fig. 7. Salt Lake City, Utah LDS Temple.

Fig. 8. Salt Lake Temple Moon Stones demonstrating the lunar cycle.

Fig. 9. Salt Lake Temple Earth Stone.

Fig. 10. Salt Lake Temple Sun Stones, Saturn Stones, Star Stones, and Big Dipper.

Fig. 11. Provo, Utah LDS Temple. The lower portion is designed to remind of a cloud, and the spire (lit at night), a pillar of fire.

What then of the cosmological imagery of the celestial, terrestrial, and telestial kingdoms? What then, of the panoramic view of God's creations and the place of mankind among them? They are found inside the temple, in the rooms where the temple ceremonies are actually performed. These rooms are designed in a pattern of upward ascent and increasing light, symbolizing the divine ascent through telestial, terrestrial, and celestial gradations of glory.[50]

They tell the story of God's plan for humankind, and thus situate the LDS worshipper within that plan. According to Brigham Young,

'Your [temple] endowment is, to receive all those ordinances in the house of the Lord, which are necessary for you, after you have departed this life, to

---

[50] For indoor, copyrighted images of LDS temple rooms, including the 'celestial room', an 'ordinance room' or creation room, a 'sealing' or marriage room, and others, see 'Inside the Temple', available at https://www.lds.org/church/temples/why-we-build-temples/inside-the-temple?lang=eng [accessed 30 November 2014].

enable you to walk back to the presence of the Father, passing the angels who stand as sentinels, being enabled to give them the key words, the signs and tokens, pertaining to the holy Priesthood, and gain your eternal exaltation in spite of earth and hell'.[51]

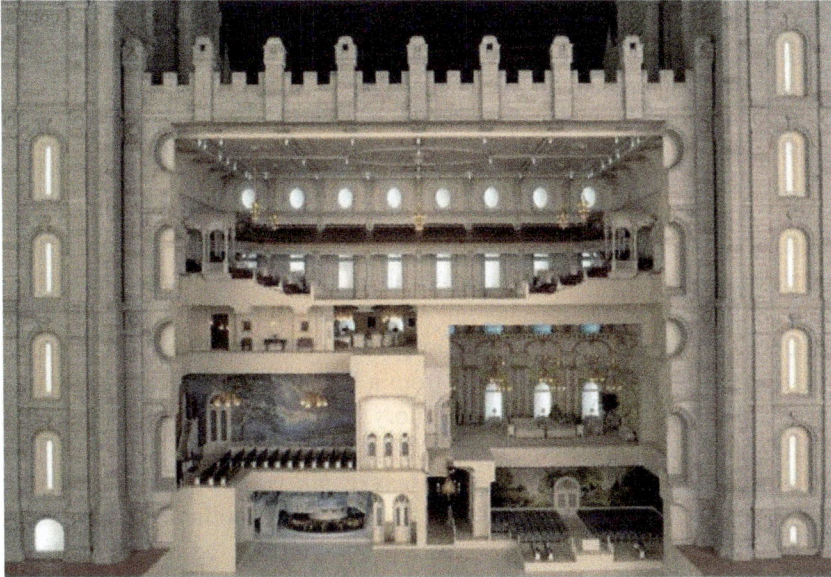

Fig. 12. Model of Salt Lake Temple located in Visitor's Center. Demonstrates the symbolic divine ascent from creation and into the Celestial room.

Solomon's temple was designed to symbolically recreate a journey back into the Garden of Eden where Adam and Eve were originally created and had stood in God's presence, and from which they had fallen.[52] As has been seen, the biblical creation and fall account were seen by Moses and Abraham in vision according to LDS scriptures. These accounts depict God's plan for his children, how those children came to arrive on earth in a fallen condition, and how they can return into God's presence. According to a current LDS apostle, 'The purposes of the Creation, the Fall, and the Atonement all converge on the sacred work done in temples of The Church

---

[51] John A. Widtsoe, ed., *Discourses of Brigham Young* (Salt Lake City, UT: Deseret Book, 1954), p. 416.

[52] See Randall Price, *Rose Guide to the Temple* (Torrance, CA: Bristol Works, Inc., 2012), p. 8–9.

of Jesus Christ of Latter-day Saints'.[53] Another church apostle stated the connection between LDS temples and the creation and fall accounts even more clearly, 'Our three accounts of the Creation are the Mosaic, the Abrahamic, and the one presented in the temples'.[54] A page from the official LDS website describes temple ordinances in this way,

> In ordinance rooms an overview is given of God's plan for His children. Latter-day Saints learn of their premortal and mortal lives, the creation of the world and the Fall of man, the central role of Jesus Christ as the Redeemer of all God's children, and the blessings they can receive in the next life.[55]

The concepts of creation, fall, and the return to God's presence also connect with theological ideas presented in the Egyptian facsimiles that were tied by Smith to the Book of Abraham. Just as the primary concern of Genesis after the creation and fall is the establishment of covenants, so Latter-day Saints see the temple as a place of covenant, in which they can progress through the grace of God from a fallen, lowly state, back through various degrees of progression and glory, and finally into the presence of God.[56] They thus participate in the grand cosmology that had been introduced in their scripture texts. Heaven meets earth; the worshippers enter the sacred 'mount' in order to rise above the pollution of the word, see a panoramic view of God's creations, be empowered and reminded of their place among God's creations, and enter symbolically into God's presence, before they return to the chaotic world in which they live; a people is prepared by covenant to become Zion-like and to welcome Enoch's city to the earth in connection with Jesus's second coming.

By way of bringing this paper full circle and back to its title, it is relevant to note that there is only one room that is considered higher than the celestial room in the temple. The marriage or sealing rooms are connected to the celestial room and provide a final step in progression. At the place where heaven meets earth, LDS temple worshippers are 'sealed' or bound or connected to God and to each other through sacred covenant and ritual.[57] From a study of its primary sources and scriptures, LDS cosmology provides a fascinating view of the marriage of heaven and earth.

---

[53] Russell M. Nelson, 'The Atonement', *Ensign* (November 1996): p. 33.

[54] Bruce R. McConkie, 'Christ and the Creation', *Ensign* (June 1982): p. 14.

[55] 'Inside the Temple'.

[56] 'Inside the Temple'.

[57] 'Inside the Temple'; see also Nelson, 'The Atonement', p. 33.

Some Remarks on the Sky in the
Ancient Egyptian *Pyramid Texts*

Joanna Popielska-Grzybowska

Abstract: This paper will scrutinise, with reference to contextual arguments, a linguistic image of the sky as pictured by the Egyptians in the world's oldest religious texts – the *Pyramid Texts*. This is an image scattered in words and phrases, in language. It will discuss how the Egyptians described the sky, celestial spheres and stars, and what or who the sky represented according to their beliefs. How did it come into being? What name or names did the sky have? Was it personified? Furthermore, what was the sky to which the deceased pharaoh was ascending? The most significant aspect, however, is how different and original the Egyptian picturing of the sky was from those of other ancient civilisations. In addition the paper will attempt to define a linguistic and religious image of the celestial domain and cosmological correlation of elements of the created world, sky and Earth, as expressed through and in the Egyptian language of the religious texts. It is a worldview reconstituted from verbal messages and pictures recorded and preserved in language.

> The Pyramid Texts, as a whole, furnish us the oldest chapter in human thinking, preserved to us, the remotest reach in the intellectual history of man which we are now able to discern.[1]

In the very beginning there was *Nu* primaeval-waters and the creator *tm* (Atum) was drifting in it.[2] Atum was not swimming, but was only drifting

---

[1] James Henry Breasted, *Development of Religion and Thought in Ancient Egypt* (New York: Charles Scribner's Sons, 1912).

[2] Susanne Bickel, 'Creative and destructive waters,' in *L'acqua nell'antico Egitto: vita, rigenerazione, incantesimo, medicamento,* Proceedings of the First International conference for young Egyptologists, Italy, Chianciano Terme, October 15–18, 2003, ed. Alessia Amenta, Maria M. Luiselli, and Maria Novella Sordi (Rome: 'L'Erma' di Bretschneider, 2005): pp.191–200; Rolf Grieshammer, 'Nun' (1982), in *Lexikon der Ägyptologie*, ed. Wolfgang Helck and Eberhard Otto (Wiesbaden: Otto Harrassowitz, 1982–1992), Vol. IV, col. 534–535; Peter Kaplony, 'Wasser' (1992), in *Lexikon der Ägyptologie*, ed. Wolfgang Helck and Eberhard Otto (Wiesbaden: Otto Harrassowitz, 1982–1992), Vol. VII, col. 16–44; Holger Rotsch, 'The primeval ocean Nun and the terminology of water in ancient Egypt', in

---

Joanna Popielska-Grzybowska, 'Some Remarks on the Sky in the Ancient Egyptian *Pyramid Texts*', *The Marriage of Heaven and Earth*, a special issue of *Culture and Cosmos*, Vol. 20, nos. 1 and 2, 2016, pp. 113–28.
www.CultureAndCosmos.org

in the water. When he felt the will to act, he created a primeval hill. He stood on it and brought to life the first pair of gods:

§ 1652a *tm xprr qA.n.k m qAA*
§ 1652b *wbn.n.k m bnbn m Hwt-bnw m jwnw*
§ 1652c *jSS.n.k m Sw tfn.n.k m tfnt*
§ 1653a *d.n.k awj.k HA.sn m awj kA wn kA.k jm.sn*

O Atum-who-is-coming-into-being. When you became high as the mound, you rose up as the benben in the Mansion of the bird benu in Junu, you sneezed Shu, you spat out Tefnut, and you set your arms about them as the arms of ka, that your essence might be in them.[3]

It is intriguing not only how the Egyptians described the sky, celestial spheres and stars, but also what or who the sky represented according to their beliefs. How did it come into being? What name or names did it have? Was it personified?

The first divine pair brought to life by the creator Atum was: Shu – life, air, breath and wind – and Tefnut, moisture. Shu and Tefnut gave life to Geb the Earth and Nut the Sky.

Curious as it may seem, the sky in the Egyptian language is feminine. It is extremely rare that a goddess rather than a god is a deity of the sky. There are two words used to refer to the sky: one is *pt*,[4] the sky as a physical

---

L'acqua nell'antico Egitto: vita, rigenerazione, incantesimo, medicamento, Proceedings of the First International conference for young Egyptologists, Italy, Chianciano Terme, October 15–18, 2003, ed. Alessia Amenta, Maria M. Luiselli, and Maria Novella Sordi (Rome: 'L'Erma' di Bretschneider, 2005): pp. 229–39; Renata-Gabriela Tatomir, 'Coincidentia oppositorum et conjunctio oppositorum: the mental category of water in the ancient Egyptian universe', in *L'acqua nell'antico Egitto: vita, rigenerazione, incantesimo, medicamento,* Proceedings of the First International conference for young Egyptologists, Italy, Chianciano Terme, October 15–18, 2003, ed. Alessia Amenta, Maria M. Luiselli, and Maria Novella Sordi (Rome: 'L'Erma' di Bretschneider, 2005): pp. 181–87; Joanna Popielska-Grzybowska, 'Nu, continuity and everlastingness in the Pyramid Texts', in *Proceeedings of the Seventh European Conference of Egyptologists. Egypt 2015: Perspectives of Research,* ed. Joanna Popielska-Grzybowska and Mladen Tomorad (Oxford: Archaeopress, forthcoming).
[3] *PT* 600 § 1652a–1653a. Unless otherwise stated, all translations that follow are the responsibility of the author.
[4] Adolf Erman and Herman Grapow, *Wörterbuch der Ägyptischen Sprache* (Berlin: Akademie-Verlag, 1926–1961), Vol. I, pp. 490–92; Raymond O. Faulkner, *A Concise Dictionary of Middle Egyptian* (Oxford: Griffith Institute, 2002), p. 87;

part of the Universe, and the other is, as mentioned above, the Sky Goddess Nut or the sky itself.[5] The most significant sphere of interest, however, is how different and original the Egyptian picturing of the sky was from that of other ancient civilisations. Furthermore, it is also stimulating to ponder what sky the deceased pharaoh was actually ascending to.

Many aspects of the oldest religious imagery in the Pyramid Texts have been analysed in recent decades. To commence, the paper presents Egyptian creeds and conceptions of the sky, the place of the stars in it, and first and foremost, the place which the deceased King took among the celestial bodies (after-life beliefs).

James Peter Allen studied the cosmology of the pyramid as one reflecting the 'cosmology of the night'.[6] This cosmology depicts a pharaoh's 'journey' from death to a new life and makes a parallel with the sun's journey dying in the west, unifying with Osiris in the *duat* (the region in the womb of Nut and the place underneath the Earth), and rising on the east side of the sky. This passage from west to east is mirrored in the arrangement of rooms in the pyramid. The sarcophagus chamber is the *duat*, the antechamber is the horizon Axt (*akhet*), and the corridor/passage is the sky. The dead King's body, now identified as Osiris's, lies in the most inaccessible place of the beyond (StA), in a sarcophagus designated *nwt*, namely 'oval'.

Many studies of the *Pyramid Texts* are concerned with astral[7] and solar[8] religions. Raymond O. Faulkner, for instance, showed a relationship

---

Rainer Hannig, *Die Sprache der Pharaonen. Großes Handwörterbuch Ägyptisch – Deutsch (2800-950)* (Mainz: Philip von Zabern, 1995), p. 269; Rainer Hannig, *Ägyptisches Wörterbuch*, Vol. I. *Altes Reich und Erste Zwischenzeit* (Mainz: Philip von Zabern, 2003), pp. 431–36; Rainer Hannig, *Ägyptisches Wörterbuch*, Vol. II. *Mittleres Reich und Zweite Zwischenzeit* (Mainz: Philip von Zabern, 2006), pp. 831–46.

[5] Erman and Grapow, *Wörterbuch der Ägyptischen Sprache*, pp. 214–15; Faulkner, *Concise Dictionary*, p. 127; Rainer Hannig, *Die Sprache der Pharaonen*, p. 396; Hannig, *Ägyptisches Wörterbuch*, Vol. I, p. 600; Hannig, *Ägyptisches Wörterbuch*, Vol. II, p. 1208.

[6] James P. Allen, 'The Cosmology of the Pyramid Texts', in *Religion and Philosophy in Ancient Egypt*, ed. William K. Simpson, *Yale Egyptological Studies* 3 (1989): pp. 1–28; James P. Allen, 'Reading a Pyramid', in *Hommages à Jean Leclant*, ed. Catherine Berger, G. Clerc, and N. Grimal, *BdÉ* 106/1 (Le Caire: IFAO, 1994), pp. 5–28.

[7] Raymond O. Faulkner, 'The King and the Star-Religion in the Pyramid Texts', *Journal of Near Eastern Studies* 25 (1966): pp. 153–61. For astronomical concepts

between the King and the Circumpolar Stars, Orion, Sirius, the Morning Star and the Lone Star, which – according to Faulkner – is identified with the pharaoh.[9]

Astral religion, prevalent initially, subsequently made gradual steps to make more room for the solar religion concentrated around the cult of Ra. To some extent religious notions were modulated as regards, for instance, the ways in which the monarch more willingly reached the sky, namely going up on sunbeams (see *PT* spell 508 and *PT* spell 523). Earthly life was inextricably linked to the sun god, as discussed by Hartwig Altenmüller.[10]

In contrast, Rudolf Anthes proposed that raising Ra to the dignity of the ruling god might have caused this change in justification.[11] Consequently, 'new beliefs' joined the existing faith in Atum the creator, Horus and other deities building pedigree creating the so-called Ennead from Heliopolis.[12] As regards beliefs concerning the sky, they were very picturesque and the

---

in the Pyramid Texts, see Bernadette Brady, 'The Egyptian ascension mythology of the Old Kingdom and the phenomenon of star phases', in *Current Research in Egyptology 2011. Proceedings of the Twelfth Annual Symposium*, ed. Heba Abd El Gawad, Nathalie Andrews, Maria Correas-Amador, Veronica Tamorri and James Taylor (Barnsley: Oxbow Books, 2012); Bernadette Brady, *Star phases: the naked-eye astronomy of the Old Kingdom Pyramid Texts* (forthcoming, personal communication with the author). Rolf Krauss, *Astronomische Konzepte und Jenseitsvorstellungen in den Pyramidentexten* (Wiesbaden: Ägyptologische Abhandlung Band 59, 1997); James Lull and Juan Antonio Belmonte, 'A Firmament Above Thebes: Uncovering the Constellations of Ancient Egyptians', *Journal for the History of Astronomy* 37 (2006): pp. 373–92; James Lull and Juan Antonio Belmonte. 'The Constellations of Ancient Egypt', in *In Search of Cosmic Order, Selected Essays on Egyptian Archaeoastronomy*, ed. Juan Antonio Belmonte and Mosalem Shaltout (Cairo: Supreme Council of Antiquities Press, 2009): pp. 153–92.

[8] See for instance: Hartwig Altenmüller, 'Aspekte des Sonnenlaufes in den Pyramidentexten', in *Hommages à François Daumas* (Montpellier: Université de Montpellier, 1986), pp. 1–15; James P. Allen, *Genesis in Egypt. The Philosophy of Ancient Egyptian Creation Accounts* (New Haven, CT: Yale Egyptological Studies, 1988).

[9] Faulkner, 'The King and the Star-Religion in the Pyramid Texts', pp. 160–61.

[10] Altenmüller, 'Aspekte des Sonnenlaufes'.

[11] Rudolf Anthes, 'Atum, Nefertem und die Kosmogonien von Heliopolis: in Versuch', *Zeitschrift für Ägyptlogische Sprache* 82 (1957): pp. 3–4.

[12] Joanna Popielska-Grzybowska, 'O Osiris Nemtiemzaf Merenre, you are the essence of all the gods', in *The Pyramid Texts as a Source of Topoi in the Coffin Texts*, Menes, (Berlin: Harrassowitz, forthcoming), p. 11–12.

imagery was truly very rich. Like the Earth, the celestial world is an expanse with cardinal directions and limits, which can be circumnavigated, and has a doorway on each side.

The main feature of the sky pt in the *Pyramid Texts* is water.[13] One passage refers directly to 'the waters... that are in the sky'(*PT* spell 685 § 2063a). The watery nature is implicit in verbs that are used to express the journey across the sky, for example: nmj, traverse (with a boat determinative); xnz, travel; Xnj, row; and DAj, cross. Probably the shortest sky description is included in James P. Allen's publication:

> ... indications from the Pyramid Texts suggest an early image of the celestial domain as an expanse (pDwt) of water (bjA, qbHw) above the Earth (Hrt), whose shores (jdbw pt) consist of marshland (sxt jArw, sxt Htp) with canals (mrw ptrw (ptrtj)) and lakes (Sjw), bordered perhaps by desert (wart, jzkn).[14]

The sky can be seen as a harbour for gods, including the sun and also Geb, the god of the Earth (incredible as it may seem), Nut, Osiris, Horus (the chief of the spirits, Horus of dAt, five times identified with the Morning Star) and birds, *kas*, *akhs* and of course stars – stars in general, but 'imperishable' ones in particular: 'You shall set me to be a magistrate among the spirits, the Imperishable Stars in the north of the sky...'.[15] The Imperishable Stars (jxmw-sk), accompany Sun-god in his Daily Barge. There are also Unwearying Stars (jxmw-wrD), named only twice in the *Pyramid Texts*, which accompany the Sun-god in his Night Barge.

§ 2172c hAjj NN m wjA ra Hr jdbw nw Sj-n-xA
§ 2173a Xn.t NN jn j.xmw-wrD
§ 2173b wD NN mdw n j.xmw-sk
The King shall go aboard the boat like Ra on the banks of the Winding Waterway, the King can be rowed by the Unwearying Stars and can give orders to the Imperishable Stars...[16]

We know that the Ancient Egyptians were familiar with observing the stars in order to determine times and seasons. They also inscribed star-maps and

---

[13] Joanna Popielska-Grzybowska, 'Contexts of Appearance of Water in the *Pyramid Texts*. An Introduction', *Études et Travaux* 29 (2015): pp. 157–67.
[14] Allen, 'The Cosmology of the Pyramid Texts', p. 9.
[15] *PT* spell 519 § 1220.
[16] *PT* spell 697 §§ 2172c–73b.

tables in their coffins and tombs (the latter can be observed at least from the Middle Kingdom).[17] Moreover, the Egyptians noticed their special characteristics:

§ 567a j jAt wrt
§ 567b sT.s wAD Szmt mfkAt sbAw

O you (goddess) who stride very wide
As she sows the green, the malachite, the turquoise of the stars.[18]

All stars to some extant are connected or collated to the King. However, all other stars but *Sah* (sAH, Orion) and *Sothis* are very prevalent in the Egyptian *Pyramid Texts*.

Sah is described as 'long of leg, wide of stride, who is at the head of Upper Egypt',[19] and 'father of gods'.[20] In religious texts *Sah* is related to Osiris, who ascends, after his death, to heaven as *Sah*.

Sah is also described as the brother and assistant of the dead, sometimes in league with *Sothis*, which leads the deceased King to the sky:

§ 822a xmtnw.Tn pj spdt wabt swt
§ 822b stt sSmw.s Tn jr wAwt nfrt jmt pt
§ 822c m sxt jArw

Your third is Sothis, pure of places; she will lead you in the Field of Reeds to the perfect routes in the sky.[21]

Sothis – the Greek name for Egyptian spdt (which means 'sharp', 'pointed') – is identified with the star called Sirius and this is generally accepted. Sothis is 'she who makes the year' (or yearly offerings). But if the King is not admitted to heaven – the new year would not be born, time would cease:

§ 1436c xsf w mswt sAH
§ 1436d xsf.k A/w jw NN jr bw ntj.k jm
§ 1437a xsf w mswt spdt

The birth of Sothis shall be prevented if you prevent the King from coming to the place where you are.[22]

---

[17] See Lull and Belmonte, 'A Firmament Above Thebes', pp. 373–92; Lull and Belmonte, 'The Constellations of Ancient Egypt', pp. 153–92.
[18] *PT* spell 350 § 567a-b.
[19] *PT* spell 477 § 959e.
[20] *PT* spells 273–74 § 408c.
[21] *PT* spell 442 § 822.

Sothis and another very significant celestial object, the Morning Star (sbA dwA, nTr dwA), apparently rest at will, they are free movers. They control time – time does not control them. However all stars differ; they are all imperishable by nature.

§ 1080a   sA.j jr sA n nTrw jpw mHtjw pt
§ 1080b   j.xmw-sk nj sk.j
§ 1080c   j.xmw-bdS nj bdS.j
§ 1080d   j.xmw-znjw nj znjw.j

I am back to the back of those gods northern of the sky – the Imperishable Stars and I will not perish,
(they) who cannot grow fatigued – I will not grow fatigued,
(they) who cannot pass away – I will not pass away.[23]

Or even more explicitly:

§ 1468d   tmjw mt n mt nb nj mt NN n mt nb
§ 1469a   j.xm-sk pw NN ...

O you who die not because of anyone dead, the King will not die because of anyone dead, for the King is an Imperishable Star...[24]

In this review of Egyptian religious notions we cannot omit the King,[25] who 'is not for the Earth: the King is for the sky'.[26] His destiny in the *Pyramid Texts* is to 'go forth to the sky among the Imperishable Stars'[27] and 'go around the sky like the sun'.[28] Immutability is the hallmark of the King's new existence, which is expressed in his identification with the 'Imperishable Stars':

§ 1080a sA.j jr sA n nTrw jpw mHtjw pt
§ 1080b j.xmw-sk nj sk.j

---

[22] *PT* spell 569§ 1436c–1437a.
[23] *PT* spell 503 §1080a–d.
[24] *PT* spell 571 §1468 d–1469a.
[25] On more aspects of the ascension of the King to the sky, see Whitney M. Davis, 'The Ascension-Myth in the Pyramid Texts', *Journal of Near Eastern Studies* 36 (1977): pp. 161–79, and the discussion of this interpretation in Joanna Popielska-Grzybowska, '"I am Life" – Linguistic View of the Egyptian Creator in the Pyramid Texts', forthcoming.
[26] *PT* spell 467 § 890d.
[27] *PT* spell 509 § 1123a.
[28] *PT* spell 218 § 130d.

§ 1080c j.xmw-bdS nj bdS.j
§ 1080d j.xmw-z njw njznjw.j[29]

The King 'will not fall to the Earth from the sky' and those who lead him to the sky 'cannot drop him to the Earth'.[30] The deceased King becomes a star (different names are involved):

The King is a sHd – star[31]
He is the nxx – star of the Lower Sky (njwt)[32]
Be long-enduring, O King, on the underside of the sky with the Beautiful Star[33]

Then the King assumes authority over the stars who kowtow before him:

§ 537a NN pw Dsr jm xnt Tzz HAt
§ 537b sbA kssw n.f nTrw sdAw n.f psDtj
§ 537c jn Drt NN wTz.s sw
The King is a holy one with sweeping forefront, lifted brow, a star to whom the gods bow, at whom the Two Enneads tremble, and it is my hand which will raise me up.[34]

The Circumpolar stars assist the King, who afterwards becomes one of them and then assumes authority over them: 'The King will not die because of any dead, for the King is a Circumpolar Star'.[35]

§ 997a nDrw n.f jt n NN tm a n NN
§ 997b wd.f NN m xnt nTrw jpf
§ 997c sbqw sAAjw j.tmw skj
The father of the King, Atum, takes the King's hand and sets him at the head of those gods who are excellent and wise, the Circumpolar Stars.[36]

We cannot finish without mentioning the Morning Star. It is often addressed in prayers in order to help the dead King to run or to ascend to

---

[29] *PT* spell 503 §1080a–d, see translation on p. 11.
[30] *PT* spell 548 § 1345a, *PT* spell 697 § 2171b.
[31] *PT* spell 571 § 1470, *PT* spell 585 §1583.
[32] *PT* spell 262 § 332.
[33] *PT* spell 684 § 2061.
[34] *PT* spell 328 § 537.
[35] *PT* spell 571 § 1468–1469.
[36] *PT* spell 480 § 997.

heaven. Even though all stars are mighty, everything, in fact, seems to serve the Lone Star (sbA watj) – the King!

The King as a star must have free will and freedom of motion to be able to obey the command:

§ 1295a wD.n jnpw xntj zH-nTr hAjj.k m sbA m nTr dwA

Anubis, chief of the zH-nTr has commanded that you come as a star, as god of the morning the Morning Star.[37]

Raymond O. Faulkner presumes that the Morning Star is Phosphorus-Venus as seen at dawn and suggests that the Lone Star is Hesperus-Venus as seen just after sunset.[38] The Lone Star is seen when no other is visible, which is very significant when one assumes the King to be the Lone Star. The only other planet which could be considered to be the Lone Star, according to Faulkner, because it can be visible when others are not, is Jupiter – but it is not as bright as Venus.[39]

One should also mention the fields described in the Pyramid Texts. The two main ones are: the Field of Reeds (sxt jArw) – Morning Star harbour[40] – and the Field of Offering (sxt Htp) (which lies north of the Field of Reeds), the abode of the Imperishable Stars. Some scholars insist that they are identical, for example Herman Kees,[41] but others argue that they are in no way the same.[42] Opinion is truly divided on this issue. To support both of these assumptions one can quote the *Pyramid Texts* § 1205c-d, translated by Jean Leclant: 'Le Champ des roseaux s'inonde, le

---

[37] *PT* spell 536 § 1295a.

[38] Faulkner, 'The King and the Star-Religion in the Pyramid Texts', pp. 153–61.

[39] Faulkner, 'The King and the Star-Religion in the Pyramid Texts', pp. 153–61.

[40] *PT* spell 437 § 805a.

[41] Hermann Kees, 'Earu-Gefilde', in Hans Bonnet, ed., *Reallexikon der Ägyptischen Religionsgeschichte* (Berlin: 1952): p. 161–62.

[42] Raymond Weill, *Le champ des roseaux et le champ des offrandes dans la religion funéraire et la religion générale* (Paris: Librairie orientaliste Paul Geuthner, 1936); Abbas Bayoumi, *Autour du champ des souchets et du champ des offrandes* (Kairo: Service Des Antiquites De L' Egypte, 1940); Samuel A.B. Mercer, *The Pyramid Texts* (New York: Longmans, 1952): Vol. IV, pp. 65–68; Jean Leclant, 'Earu-Gefilde', in Wolfgang Helck and Eberhard Otto, eds., *Lexikon der Ägyptologie* (Wiesbaden: Otto Harrassowitz, 1975): vol. 1, col. 1156–60.

Champ des offrandes se remplit d'eau'.[43] This passage can be treated as evidence of their sameness or of their difference.

Besides the fields, James P. Allen associates msqt sHdw 'beaten path of stars' – perhaps the contemporary Milky Way – with the only body of land in the sky. However, as he admits himself, apart from the determinative of a land strip or a desert, it can also have (see *PT* spell 254 § 279d) a determinative which may suggest that it could be crossed by boat.[44]

When, however, it comes to describing and scrutinising the roles of Nut the sky, it turns out to be a much more complex issue, one that was thoroughly scrutinised by Nils Billing.[45] In the *Pyramid Texts* Nut is the mother of the King:

§ 777a nwt pSS Tn Hr zA.T wsjr ppj
§ 777b sdx.T sw m a stS Snm sw nwt
§ 777c jw.n.T sdx.T   zA.Tjw.n.(T)   js Snm.(T) wr pn

Nut, spread yourself over your son, Osiris Pepi,
and conceal him from Seth.
Join him, Nut, who comes to you, and conceal your son
as he who comes to you: (you) shall join this great one.[46]

Nut is the sky full of the stars, but the pharaoh is the Lone Star 'at Nut's shoulder':

§ 250a j n.T wnjs pn nwt jn.T wnjs pn nwt
§ 250b qmA.n.f jt.f   r tA fx.n.f Hrw m xt.f
§ 250c rd DnHwj.f m bjk j.Swtj m gmHsw
§ 250d jn.n sw bA.f Htm.n sw HkAw.f
§ 251a wp.k st.k m pt m ab sbAw nw pt
§ 251b n Twt js sbA watj rmn nwt Hw A mA.k Hr tpj wsjr
§ 251c wD.f mdw n Axjw Twt aHa.tj Hr.t r.f
§ 251d nj Tw jm.sn nj wnn.k jm.sn

This Unis has come to you, Nut; this Unis has come to you, Nut,

---

[43] Jean Leclant, 'Earu-Gefilde', in *Lexikon der Ägyptologie*, ed. Wolfgang Helck and Eberhard Otto (Wiesbaden: Otto Harrassowitz, 1975): Vol. 1, col. 1156.

[44] Allen, 'The Cosmology of the Pyramid Texts', p. 7.

[45] Nils Billing, *Nut. The Goddess of Life in Text and Iconography* (Uppsala: Uppsala University, 2002).

[46] *PT* spell 427 § 777.

Having given his father to the Earth, having left Horus behind him,
Having grown wings as a falcon, feathered as a hawk,
His ba having brought him, his magic having provided him.
Open your place in the sky among the stars of the sky,
for you are the Lone Star at Nut's shoulder. May you look upon the head of
Osiris as he gives orders to the akhs, while you yourself stand far from him:
you are not among them and you should not be among them.[47]

Nut seems to be so powerful and decisive that if she supports the King he shall not die:

§ 779a Dd-mdw jn gbb
§ 779a nwt Ax.n.T
§ 779b sxm.n.T m Xt mjw.T Ttfnt nj ms.t.T
§ 779c Snm.T ppj pn nj mt.f
Words to say by Geb:
Nut, as you became luminous
And had power in the belly of your mother Tefnut before you were born,
may you join this Pepi, and he will not die.[48]

Another very significant feature of the sky in the *Pyramid Texts* is that it is the place that was the destination of the journey of the deceased pharaoh. He may reach the sky by very different means, but always is moving upwards, thus ascending:

§ 365a sq.t n.f tA rdw r pt pr.f jm r pt
§ 365b prr.f Hr Htj n jdt wrt
§ 366a j.pA NN pn m Apd xnn.f m xprr
§ 366b j.pA.f m Apd xnn.f m xprr
§ 366c m nst Swt jmt wjA.k ra
A stairway to the sky is set up for him that he may mount upon it to the sky –
and he will ascend on the smoke of the great censing.
This Unis will fly up as a bird and alight as a beetle – he does flies up as a
bird and alights as a beetle on the empty throne which is in your bark, Ra.[49]

The pharaoh is predestined to the sky but his body shall stay on Earth:

§ 474a Axj jr pt SAt jr tA
§ 474b Szpt rmT qrs.sn

---

[47] *PT* spell 245.
[48] *PT* 429.
[49] *PT* spell 267 § 365–66.

§ 474c xA.s m t xA.s m Hnqt Hr wdHw n xnt jmntjw

Akh to the sky! Body to the Earth!
What people receive when they have been buried, their thousand of bread and
their thousand of beer, it is from the offering table of the Foremost of the
Westerners.[50]

Sometimes also occurrences taking place in the sky were described as follows:

§ 304a Xnnw m pt
§ 304b mA.(n).n mAt jn.sn nTrw pAwtjw
§ 304c psDt Hrw m jAxw
§ 304d snhd n.f nbw jrw
§ 304e pSr n.f psDtj tm
§ 305a Hms r.f m st nb tm jT wnjs pt pSn.f bjA.s
§ 305b sSmw wnjs wAwt n xpr
§ 306a Htp wnjs m anx m jmnt Sms sw dAtjw
§ 306b psD wnjs mAj.j m jAbt
§ 306c jw.t n.f wp Xnnw m ksw
§ 306d snhd.n wnjs nTrw smsw r wr
§ 306e jn sw sxm m st.f
§ 307a jT wnjs Hw jn.t n.f nHH
§ 307b smn.t n.f sjA jr rdwj.f
§ 307c hjj n wnjs jT.n.f Axt

Confusion in the sky!
'We saw something new,' say the primeval gods.
Ennead, Horus is in the sunlight. Let those who have forms terrorise for him,
let Dual Ennead of Atum serve him as he sits on the seat of the Lord of All.
Unis will acquire the sky and open its plain;[51]
Unis will guide the paths of This-who-is-coming-into-being.
When Unis rests in life in the west, those of the Duat will guide him;
When Unis shines again in the east,
he who parted the two contestants will come to him kowtowing.
The gods will serve Unis, since he is older than the great one
and power belongs to him in his place.
Unis will assume Authoritative Utterance, Eternal Continuity will be brought
to him, and Perception shall be made enduring for him at his feet.
Rejoice for Unis, for he has possessed the Akhet![52]

---

[50] *PT* spell 305 § 474.
[51] See also Allen, *The Ancient Egyptian Pyramid Texts*, p. 45.
[52] *PT* spell 257.

The King was then the one who controlled all the powers on the Earth, but also in the sky and he himself was the power over the powers there. Moreover, the pharaoh could have multiple identities: he could be an embodiment of the sun, the Lone Star, one of the Imperishable Stars, or Osiris, and was to reach his final stage of existence as the creator god Atum. The monarch's lot is not to enter into the Earth, because his destination is to live in the sky.

§ 308a wsjr pw wnjs m zzw
§ 308b bwt.f pw tA nj aq wnjs m gbb
§ 308c Htm.f qd.f m Hwt.f tp tA
§ 308d sD/srwD qsw.f dr sDbw.f
§ 308e wab.n wnjs m jrt Hrw dr sDb.f m Drtj wsjr
§ 308f sfx.n wnjs rDw.f m gsA jr tA
§ 309a jn snt.f nbt p rmt sw
§ 309b jw wnjs r pt jw wnjs r pt m TAw m TAw
§ 309c nj xm.f nj xmwt.f jm.f
§ 309d nj Hmsw wnjs m DDAt nTr
§ 309e wnjs pj Hr wa.f smsw nTrw
§ 310a jw pAD.f jr Hr Hna ra
§ 310b jw aAbt.f m nw
§ 310c wnjs pj nnw
§ 310d Sm.f/j.Sm wnjs (Hna ra) jw.f/jw wnjs Hna ra
§ 310e zxn.f Hwwt.f

Unis is Osiris in a dust-devil.
The Earth is his abomination: he shall not enter into Geb.
He will annihilate his sleep broken in his mansion on Earth,
his bones are strengthened and his obstacles removed.
Unis has become pure through the eye of Horus,
his obstacle has been removed by the two kites of Osiris,
and Unis has offered his efflux to the Earth in Qus.
His sister, the Lady of Pe, is the one who mourned for him:
Unis is off to the sky, Unis is off to the sky, in the wind, in the wind. He shall not be debarred, and there is no-one who will be rejected from him. He shall not sit in the Tribunal of the god.
Unis is the one who is on his own, the oldest of the gods: his offering of bread is up above with Ra, his provisions are from Nu.
Unis is the one who goes to and fro, going and coming with Ra and owning his mansions.[53]

---

[53] *PT* spell 258.

Another occurrence was pictured in the *Pyramid Texts*. This one involves all the characters portrayed in this text:

§ 458a sbS pt anx spd(t) n wnjs
    js anx zA spdt
§ 458b wab.n n.f psDtj
§ 458c m msxtjw j.xm-sk
§ 458d nj sk pr wnjs r pt nj Htm nst wnjs jrt tA
§ 459a dx r.sn rmT pAjj r.sn nTrw
§ 459b spA.n spdt wnjs jr pt m ab snw.f nTrw
§ 459c kf.n nwt rmnwj.s n wnjs
§ 460a qfn.n snj bAwj xntw bAw jwnw Xr tpj ra
§ 460b sDr jr.sn nn n rmwt nTr

The sky has been cleansed and Sothis lives, for Unis is the living one, son of Sothis,
for whom the Dual Ennead have purified the Imperishable in Meskhetyu.
House of Unis will not perish for the sky, throne of Unis will not finish for the Earth.
People have hidden, the gods have flown away, for Sothis has made Unis fly to the sky amidst his brothers, the gods.
Nut has uncovered her arms for Unis;
the two leading bas of the bas of Junu, who spent the night making this bewailing of the god, have bowed down at the head of Ra.[54]

When the pharaoh is on his way to take his place in the sky every creature rejoices. This fact depicts the overwhelming joy of life, which was inextricably linked to ancient Egyptian culture, even though – judging by all appearance and solicitude when practices regarding funerary rituals are concerned – one may think of this culture as a culture of death:[55]

§ 476a Dd mdw nfr w a mAw Htp w A ptr jn.sn jn nTr
§ 476b prt   r.f nTr pn jr pt prt r.f jt wnjs jr pt
§ 477a bAw.f tp.f Sat.f r gswj.f
§ 477b HkAw.f tp rdwj.f
§ 477c jr.n n.f gbb mr qd jrjj n.f jm

---

[54] *PT* spell 302 § 458–460b.
[55] Joanna Popielska-Grzybowska, "O królu! Obróć się! O królu! Krzycz! Krzycz! Dzień po dniu, noc w noc' – egipski zachwyt nad życiem', in *Przez granice czasu. Księga jubileuszowa poświęcona Profesorowi Jerzemu Gąssowskiemu*, ed. A. Buko, W. Duczko, J. Popielska-Grzybowska, B. Jurkiewicz, and A. Stawiska (Pułtusk: Wydawnictwo Akademii Humanistycznej, 2008), pp. 373–79.

§ 478a j n.f nTrw bAw p nTrw bAw nxnw nTrw jrw pt
nTrw jrw tA
§ 478b jr.sn wTzw n wnjs Hr awj.sn
§ 479a pr.k r.k wnjs jr pt j.Aq Hr.s m rn.s pw n mAqt
§ 479b rD pt n wnjs rD n.f tA jn tm
§ 480a mdw Hr.s pw gbb
§ 480b jAwt jAt jA(w)t Hrw jA(w)t stS
§ 480c sxwt jArw dwA.sn Tw
§ 480d m rn.k pw n dwA spd js Xrj ksbwt.f

Words to say: 'How beautiful is the sight, how pleasing the vision,'
*Say they, namely the gods, of this god's ascending to the sky, of Unis's
ascending to the sky, with his bas upon him, his severity at his sides, his magic
at his feet.*
Geb has acted for him likewise it was done for him.
*There have come to him the gods, the bas of Pe and the gods, the bas of Nekhen,
the gods who belong to the sky and the gods who belong to the Earth, that they
might make for Unis lifting on their arms.*
*You Unis shall ascend to the sky and come up on it in this its identity of the
ladder.*
'The sky shall be given to Unis and the Earth shall be given to him,' says Atum.
*Geb is the one who speaks for it: 'The mounds that are adored, that Horus
adores, that Seth adores, and the Marshes of Reeds shall worship you in your
identity of the Morning God, like Seped under his kesebut-trees.*[56]

It is thought provoking that in many contexts where the sky and Nut
appear, the Earth god Geb is also evoked. Sometimes, however, as in the
case of the passage quoted below, the King is identified with the Earth god
Geb, husband of Nut, the sky to give life to the world, to all the creatures,
gods and human beings, animals and plants.

§ 783a n pnd.n Tm m gbb m rn.T n pt
§ 783b zmA.n n.T gbb tA r Dr.f m bw nb

*As Geb has I fertilised you in your identity of the sky, I have consolidated the
entire land for you in every place.*[57]

Furthermore, the Egyptian sky consisted of the upper sky and the under-
sky, but the sky was not one of the first creatures which came into being or
were brought to life by the creator. The pharaoh was born before anything
else came into being.

---

[56] *PT* spell 306 § 476–80.
[57] *PT* 433.

§ 1466a jwr mjwt nt NN jm.f jm nwt
§ 1466b ms NN pn jn jt.f tm
§ 1466c nj xpr.t pt nj xpr.t tA
§ 1466d nj xpr.t rmT nj ms.t nTrw nj xpr.t mt

The King's mother was pregnant with him, who is in the under-sky, the King was fashioned by his father Atum before the sky came into being, before the Earth came into being, before people came into being, before the gods were born, before death came into being.[58]

How beautiful and alluring, then, were the celestial domains according to the ancient Egyptians? The Egyptian knew well how to enjoy the beauty of nature and life, something that also differentiated them significantly from many other ancient civilisations.

However, although it is extremely difficult to discern, it may be concluded tentatively that the imagination which enabled the Egyptians to create the celestial world was also capable of employing figurative images. Let us then give the final words to the ancient Egyptians as they recorded in the Pyramid Texts:

§ 782a Dd mdw aAt xpr.t m pt n sxm.n.T
§ 782b n jmjm.n.T mH.n.T bw nb m nfr.T
§ 782c tA Xr.T r Dr.f jT.n.T sw
§ 782d Sn.(n).T n.T tA xt nb m Xnw awj.T
§ 782e d.(n).T n.T ppj pn m j.xm-sk jm.T

Words to say: O something great has happened in the sky, you have achieved control, you have achieved power, and have filled every place with your beauty. The entire land is yours, for you have taken possession of it. You have enclosed the Earth and every thing within your embrace, and you have placed this Pepi as an Imperishable Star who is in you.[59]

---

[58] *PT* 571§ 1466.
[59] *PT* 432§ 782.

From the Margins to the Image of 'The Most Christian Science': Astrology and Theology from Albert the Great to Marsilio Ficino

Scott Hendrix

Abstract. In 1277 a committee headed by Stephen Tempier, the bishop of Paris, issued the famed Condemnations of 1277. Several propositions were aimed directly at astrological beliefs. Yet by the end of the fifteenth century, astrology was an accepted part of the intellectual landscape of Europe. This study argues that the thirteenth century saw tremendous controversy over the acceptance of astrology, but Albert the Great's writings set the terms for the debate in the following centuries. Albert's understanding of astrological forecasting and celestial influence played a key role in astrology's transition from being a marginal discipline, to one that European intellectuals and church leaders came to see not only as largely non-controversial but, in fact, as central to efforts to understand God's divine plan.

As Greek, Hellenistic, and Arabic learning slowly made an impact on the European intellectual landscape in the twelfth and thirteenth centuries, interest in the relationship between the heavens and the earth – and the potential to predict future events or even manipulate celestial influence – was an important, and perhaps even leading, factor in the drive to gain access to these forms of knowledge.[1] The reason for this is simple: astrology was seen as useful to a wide range of people, from princes in need of advice to physicians in need of diagnostic tools to enable better treatment of their patients. Moreover, many felt that it was the highest form of philosophical knowledge available, making astrological study and

---

[1] Richard Lemay has pointed out that translations of Abu Ma'shar's astrological work preceded most of Aristotle's corpus, arguing that interest in Abu Ma'shar – and indeed interest in astrology – was one of the primary factors driving the revival of interest in Aristotle. See Richard Lemay, *Abu Ma'shar and Latin Aristotelianism in the Twelfth Century* (Beirut: American University of Beirut, 1962), introduction.

Scott Hendrix: 'From the Margins to the Image of "The Most Christian Science": Astrology and Theology from Albert the Great to Marsilio Ficino', *The Marriage of Heaven and Earth*, a special issue of *Culture and Cosmos*, Vol. 20, nos. 1 and 2, 2016, pp. 129–46.
www.CultureAndCosmos.org

knowledge prestigious and valued.[2] True, this view did not go uncontested, but by the end of the thirteenth century one of the greatest scholars of his age, Albert the Great (1200–1280), had crafted a defense of astrology which calmed fears that astrological forecasting not only did not compromise free will, but in fact could improve the free exercise of the will. By the late fifteenth century, Albert's arguments had become commonplace and most intellectuals saw astrological study as a licit and highly prestigious pursuit. Furthermore, at a time when free will and human potential were valorized, many found the argument that astrological study could enhance the freedom of the will to be quite attractive.

There was little if any systematic practice of astrology in western Europe from the collapse of Roman power until the twelfth century. The mathematical skills, tables, and other technical elements necessary to practice this discipline had simply been lost. True, interpretations of celestial phenomenon still occurred in these early centuries. For example, in 837 Louis the Pious asked a member of his court, a man remembered simply as 'the Astronomer,' to interpret a particularly vivid comet that appeared in the Spring sky, which most now agree was Halley's comet.[3] However, the sorts of predictions that occurred from time to time in the early middle ages lacked rigorous mathematical grounding.[4]As mathematical astronomy and the associated technical skills began to be reintroduced to the West in the twelfth century, organized opposition to the discipline was slow to mount.[5] There were occasional references to church fathers such as Augustine (d. 430), who in book five of his *City of God* took astrology to task, asserting that the differing fates of twins

---

[2] This was certainly Abu Ma'shar's position, which would in turn influence many readers in Latin Christendom. See Lemay, *Abu Ma'shar and Latin Aristotelianism in the Twelfth Century*, 242. Albert the Great states that 'videmus astronomia omnes alias scientias excellere'. In English: 'I see astrology to excel all other sciences'. All translations are my own. Albertus Magnus, *De anima* in his *Opera Omnia*, August Borgnet, ed. (Paris: Vives, 1840), book I, tract. I, ch. II, p. 119.

[3] Bruce S. Eastwood, *Ordering the Heavens: Roman Astronomy and Cosmology in the Carolingian Renaissance* (Leiden: Brill, 2007), p. 151.

[4] Stephen C. McCluskey, *Astronomies and Cultures in Early Medieval Europe* (Cambridge: Cambridge University Press, 1998), pp. 188–90.

[5] Marie-Thérèsed'Alverny, 'Translations and Translators', in *Renaissance and Renewal in the Twelfth Century*, edited by Robert Louis Benson, Giles Constable, and Carol Dana Lanham, (Toronto: Medieval Academy of America, 1999), pp. 421–62, here pp. 453–55.

demonstrated astrology's lack of reliability or that the doctrine of the freedom of the will necessitated the absence of fate.[6] Sometimes attacks on astrological forecasting were more fully developed, as in the case of Moses Maimonedes (1135–1204) who saw astrology as leading toward idolatry and potentially representing a threat to free will.[7] Yet, given the fact that both Augustine and Maimonedes accepted the notion that celestial forces affected terrestrial creatures and events, the door was left wide open for those who would find its study to be too useful to pass up – which would prove to be true for a wide range of people. By the end of the twelfth century it would be all but impossible to find a physician who did not use astrology as a primary diagnostic tool or a court where an astrologer was not a principle adviser; in fact, these individuals were often one and the same, as the adviser who provided astrological castings for his prince was often also his physician.[8]

In the following century opposition to astrology would increase in direct relationship to the growth in interest in the discipline. As universities spread across Europe, an increasing number of people became acquainted

---

[6] Augustine, *De Civitate Dei,* Corpus Christianorum Series Latina 48 (Turnhout: Brepols, 1955), p. 129. 'quod fit, quod nihil umquam dicere potuerunt, cur in vita geminorum, in actionibus, in eventis, in professionibus, artibus, honoribus ceterisque rebus ad humanam vitam pertinentibus atque in ipsa morte sit plerumque tanta diversitas, ut similiores eis sint, quantum ad haec adtinet, multi extranei quam ipsi inter se gemini perexiguo temporis intervallo in nascendo separati, in conceptu autem per unum concubitum uno etiam momento seminati?' In English: 'How is it that they [astrologers] have never been able to say why in the life of twins, in actions, events, professions, fields of knowledge, honors, and other things pertaining to human life and even in death itself, there is such diversity with regard to these things many strangers are more similar to them than twins are to each other, even though when they were born they were separated by the smallest interval of time and, moreover, in conception, they were conceived by a single act of conception even in the same moment'. See also, L. C. Ferrari, '*Augustine* and *Astrology*', *Laval Theologique et Philosophique* 33 (1977): pp. 241–51.

[7] Moses Maimonides, 'Letter on Astrology', in Isadore Twersky, *A Maimonides Reader* (New York: Behrman House, 1972), pp. 463–73; Y. T. Langermann, 'Maimonides' Repudiation of Astrology', *Maimonidean Studies* 2 (1997): pp. 123–58.

[8] Lynn Townsend White, Jr., 'Medical Astrologers and Late Medieval Technology', in *Medieval Religion and Technology: Collected Essays*, edited by Lynn T. White, Jr., (Los Angeles: University of California Press, 1978): pp. 297–316.

with the tenets of astrology and its divinatory applications. After all, astrology and astronomy were seen as representing a single discipline during the middle ages, with many intellectuals using the terms interchangeably or in ways that are exactly the opposite to what a modern reader might expect.[9] Since astronomy was one of the seven liberal arts, even if astrological study was largely reserved for medical students, any of the million or more students who had attended a European university by 1500 would have become conversant in the basic tenets of astrology.[10] Furthermore, during this period first physicians and then later surgeons became professionalized, requiring certain standards of training and licensure – including a mastery of astrology and its applications.[11]

Perhaps due to its greater visibility, by the end of the thirteenth century astrology encountered intensifying opposition. The most visible and famous instances of this opposition occurred in Paris, in the context of the Condemnations of 1270 and 1277. Though neither of these lists of condemned prohibitions was directed solely at astrology, if either had been effective the study and application of astrology would have been stifled. Taken jointly, the most important concern expressed in 1270 and again in 1277 is the idea that astrological belief could compromise the freedom of the will. In 1270 two of the thirteen prohibited positions were:

---

[9] Charles Burnett, 'Astrology', in *Medieval Latin: An Introduction and Bibliographical Guide*, edited by Frank Mantello and A. G. Rigg, (Washington, DC: The Catholic University of America Press, 1999): pp. 369–82.

[10] Edward Grant, *Planets, Stars, and Orbs: The Medieval Cosmos, 1200–1687* (Cambridge: Cambridge University Press, 1996), p. 137; Edward Grant, *The Foundations of Modern Science in the Middle Ages: Their Religious, Institutional, and Intellectual Contexts* (Cambridge: Cambridge University Press, 1996), pp. 37–38, 44–45. The number he provides is 750,000 for the period from 1350 to 1500. When we extrapolate back to the end of the twelfth century, one million seems to be a conservative estimate.

[11] Roger French, *Medicine Before Science: The Business of Medicine from the Middle Ages to the Enlightenment* (Cambridge: Cambridge University Press, 2003), pp. 88–120. Contrary to what a modern person might expect, surgeons were considered to be craftsmen, rather than professionals and intellectuals, until the fourteenth century. Roger French, 'Astrology in Medical Practice', in *Practical Medicine from Salerno to the Black Death*, edited by Luis Garcia-Ballester, Roger French, Jon Arrizabalaga, and Andrew Cunningham, (Cambridge: Cambridge University Press, 1994), pp. 30–59.

(4) that all that happens here below is subject to the necessity of the heavenly bodies.

(9) that free will is a passive power, not an active one, and that it is necessarily moved by the object of desire.[13]

In 1277 the commission again headed by Bishop Stephen Tempier (d. 1279) voiced further concerns about the possibility that astrological beliefs could compromise the freedom of the will. These condemnations were much more thoroughgoing in their assault on astrological doctrines, with those relating most directly to the question of free will being:

(21) That nothing happens by chance and that all things occur due to necessity.

(161) That the effects of the stars upon free will are hidden.

(162) That our will is subject to the power of the celestial bodies.[14]

It is important to note that few or none actually held such extreme positions as those articulated in the Condemnations. Furthermore, there is no evidence of anyone being prosecuted for direct violation of the condemned propositions. Nevertheless, the condemned positions do show a deep concern about astrology – particularly in regards to its relationship to free will.

An examination of the events leading up to the Condemnations of 1270 and 1277 would take us too far afield from our present study. For now, we should simply note that they were ineffective in stopping or even slowing the study and practice of astrology in its various forms, or in accomplishing anything else for that matter.[15] What is more important to

---

[13] John F. Wippel, 'The Condemnations of 1270 and 1277 at Paris', *The Journal of Medieval and Renaissance Studies* 7, no. 2 (1977): pp. 169–201, 179; Henri Denifle and Emile Chatelain, O.P., eds., *Chartularium Universitatis Parisiensis* (Paris: Delalain, 1889), I:487. '(4) Quod omnia, que hic in inferioribus aguntur, sub sunt necessitate corporum celestium. (9) Quod liberum arbitrium est potential passiva, non activa; et quod necessitate movetur ab appetibili.'

[14] D. Piché and C. Lafleur, *La condemnation parisienne de 1277, nouvelle édition du textelatin, traduction, introduction et commentaire* (Paris: J. Vrin, 1999), pp. 74–130. For the prohibitions of 1277, see Denifle and Chatelain, O.P., eds., *Chartularium*, Vol. I, pp. 551–55.

[15] Roland Hissette, *Enquête sur les 219 articles condamnés à Paris le 7 mars 1277* (Louvain: Publications Universitaires, 1977); For the larger context, see Scott E. Hendrix, *How Albert the Great's Speculum Astronomiae Was Interpreted and Used by Four Centuries of Readers* (Lewiston: Edwin Mellen, 2010), ch. 2.

the current study is *why* these assaults on astrology were ineffective. To put it simply, a defense of astrology had already been created by one of the most important theologians of the day, a defense that would prove to be enormously influential for centuries to come.

Albert the Great was described lovingly by one of his students, Ulrich of Strasbourg, as 'a man so superior in every science, that he can fittingly be called the wonder and miracle of our time'.[16] This was not merely hyperbole, nor simply the opinion of a student overawed by his master: writing between 1266 and 1267 Roger Bacon, by no means friendly toward Albert, stated that a certain master, presumably Albert, was known as an authority in Paris on a par with Aristotle, Avicenna, and Averroes.[17] According to Bacon, Albert's reputation exceeded that which any other master had ever held during his own lifetime.[18] The honor that Albert's order bestowed upon him by naming him as Prior Provincial of Teutonia in 1254 as well as the fact that Pope Alexander IV invited Albert to his court in 1256 before personally naming him Bishop of Regensburg in 1260 also supports Bacon's statements about the strength of Albert's reputation among his contemporaries.[19]

Among other things, Albert established himself as an authority on astrology to such a degree that he was frequently called upon to address questions about the compatibility of astrological belief with orthodox Christian doctrine in the 1260s and 1270s.[20] Given Albert's importance to

---

[16] J. Daguillon, *Ulrich de Strasbourg, La 'Summa de bono'. Livre I* (Paris: J. Vrin, 1930), p. 139.

[17] Jeremiah Hackett, 'The Attitude of Roger Bacon to the *Scientia* of Albertus Magnus', *Albertus Magnus and the Sciences*, edited by James A. Weisheipl, (Toronto: Pontifical Institute of Mediaeval Studies, 1980): pp. 53–72; Roger Bacon, *Opus Tertium,* in *Opera quaedam hactenus inedita*, J. S. Brewer, ed. (London: Longman, 1859), p. 30.

[18] Bacon, *Opus Tertium*, p. 30; Hackett, 'The Attitude of Roger Bacon', pp. 53–57.

[19] Alain de Libera, *Albert le Grand et la Philosophie* (Paris: Vrin, 1990), pp. 16–17; James A. Weisheipl, 'The Life and Works of St. Albert the Great', in *Albertus Magnus and the Sciences* (Toronto: Pontifical Institute of Medieval Studies, 1980): pp. 13–53, 33.

[20] Albert the Great, 'Problemata determinata', *Opera omnia*, Jacob Weisheipl, ed. (Monasterii Westfalorum: Aschendorff, 1975), pp. 45–64. For Albert's attitudes about astrology, and his employment as an expert witness, see: J. Weisheipl, 'The *Problemata Determinata XLIII* Ascribed to Albertus Magnus (1271)', *Mediaeval Studies* 22 (1960): pp. 303–54; Fernand Van Steenberghen, 'Le 'De quindecim problematibus' d'Albert le Grand', in *Études d'histoire littéraire et doctrinale de*

this discussion, we should make sure we understand his position. We can get a clear view his attitudes about astrology from statements he supplied in response to questions Giles of Lessines (died c. 1304) posed to him. Giles was in Paris studying under Albert's star student, Thomas Aquinas (1225–1274), when he wrote to Albert, asking the elder man's opinion on a list of fifteen questions.[21] There is no doubt that this list of questions was generated by the controversy swirling around Paris in the years between 1270 and 1277. Of the fifteen questions Giles put to Albert, thirteen of them reproduced the Condemnations of Paris of 1270.[22] Several points touched directly on astrological principles, but for our present purposes, the most important point about these questions is that the response provides a platform for Albert to enunciate his defense of astrology in a succinct format.[23] This was a debate that Albert seems quite perturbed to have been called upon to address. Writing to Giles he stated:

> What those say in the third place, that the will of man wishes and chooses from necessity, no man is ever able to say such a thing unless he is deeply

---

la Scolastique médiévale offertes à Monseigneur Auguste Pelzer (Louvain: Bibliothèque de l'Université/Editions de l'Institut Supérieur de Philosophie, 1947): pp. 415–39; Albert the Great, 'De quindecim problematibus,' in *Opera omnia*, Bernhard Geyer, ed. (Monasterii Westfalorum: Aschendorff, 1975), vol. XVII, pp. 31–34.

[21] Fernand van Steenberghen, *Aristotle in the West: The origins of Latin Aristotelianism* (Louvain: Nauwelaerts, 1970), p. 212; Steenberghen, 'Le 'De quindecim problematibus' d'Albert le Grand', pp. 415–39; Albert the Great, 'De quindecim problematibus', pp. 31–34.

[22] Albert, 'De quindecim problematibus', p. 31.

[23] Albert, 'De quindecim problematibus', p. 31. The points in question are: 'III: Quod voluntas hominis ex necessitate vult et eligit. IV: Quod omnia quae in inferioribus aguntur, sub sunt necessitati corporum caelestium. IX: Quod liberum arbitrium est potentia passiva, non activa quod de necessitate movetur ab appetibili. XII: Quod humani actus non reguntur providentia dei'. In English: 'III: that the will of man wishes or chooses from necessity. IV: That all things which happen in the in the sublunar realm, are under the necessity of the heavenly bodies. IX: That free will is a passive, not an active power, that is moved from necessity by appetative desire. XII: That the actions of man are not ruled by the providence of God'.

> illiterate, because every argument as well as every school of ethics,
> whether of the Peripatetics or of the Stoics, cries out that we are the lords
> of our acts.[24]

It was uncharacteristic of Albert to react with such negativity, but we
should think about the situation from his point of view. He had already
settled these questions in works written throughout the course of his
scholarly career. Albert's writings are shot through with references to
astrology, ranging from discussions of the usefulness of various forms of
divination to simple mentions of astrological principles as support for other
arguments. Such citations are present in his earliest work, *De natura boni*
(*On the Nature of Good*), his last, the *Summa theologiae* (*Summary of
Theology*), and virtually everything he wrote in between.[25]

Despite his irritation Albert conscientiously answered Giles' questions,
for the most part anyway. Albert appealed to Hermes Trismegistus,
Aristotle, and Ptolemy to explain the differing forms of causes, as well as
Ptolemy's definition of fate. But Albert noted that 'fate' does not impose
necessity due to three causes. There can be no necessity

> because it [fate, divine influence] is not passed [to the native] directly but
> through a medium, and [fate] will be able to be impeded by its [the
> medium's] inequality [to God]. Then there is the second [reason], that it
> [fate] is not effected in natives [meaning those born under a given set of
> celestial influences] in and of itself, but through accidental characteristics;
> [thirdly] it is effected through primary qualities, which do not receive the
> powers of the stars in and of themselves, because matter –in the diversity
> and power of the matter of natives –is not able to receive the powers of the
> heavens uniformly just as they are in the heavens.[26]

---

[24] Albert, 'De quindecim problematibus', p. 35. 'Quod autem tertio dicunt, quod
voluntas hominis ex necessitate vult et eligit, numquam potuit dicere nisi homo
penitus illitteratus, quia omnis ratio et omnis ethicorum scholatam Stoicorum
quam Peripateticorum clamat nos dominos esse actuum nostrorum'.
[25] Lynn Thorndike, *History of Magic and Experimental Science* (New York:
Columbia University Press, 1923), II: pp. 584, 589. In his *Summa, pars* 1, Questio
68, Albert states that the stars govern even the souls, vegetable and sensitive, of
plants and brutes, but man is made in the image of God, except as he yields to sin
and the flesh: as such, the intellectual soul is free.
[26] Albert, 'De quindecim problematibus', p. 36: 'quod fatum, quod ex
constellatione est, necessitatem non imponit propter tres causas. Quarum unus est,
quia non immediate, sed per medium advenit, cuius inaequalitate impediri poterit;

In other words, fate is imparted through the celestial medium, rather than directly, and is therefore impeded by the inequality of the matter of the heavens in comparison to God's divine perfection. Furthermore, creatures receive this influence – their fate – in their material beings, rather than their incorporeal souls. The final result is that the fate of creatures is enacted through their 'primary qualities' – their souls. The soul, however, does not receive divine influence directly, but rather *per accidens*, imparted by corporeal impulses that are derived from the heavens. Thus, for Albert fate is a term for the combinations of influences acting to affect our willed actions and therefore plays a large role in the unfolding of future events, but it is not dispositive.[27] Even in the absence of willed acts, earthly matter lacks the perfection of heavenly quintessence, making it incapable of receiving celestial influences 'uniformly.' Albert has little patience for anyone who might disagree with this argument, stating that 'those who say otherwise are in every way ridiculous'.[28] While Albert's thinking was broadly Aristotelian, his system of thought contained strong Neoplatonic elements as well, due to his mistaken belief that the *Liber de causis*, which contained content apparently taken from the Neoplatonic philosopher

---

secunda autem, quia per accidens, sed non per se operatur in natis; operatur enim per primas qualitates, quae non per se virtutes stellarum accipiunt, in diversitate et potestate materiae natorum, quae materia uniformiter et, prout sunt in caelis, recipere non potest caelorum virtutes'. Albert expresses a similar view of the importance of the qualities of the intervening and receiving matter and how such matter can affect the transmission of influence in his *Speculum astronomiae,* included in Paola Zambelli, *The Speculum Astronomiae and its Enigma: Astrology, Theology, and Science in Albertus Magnus and his Contemporaries* (Dordrecht: Kluwer Academic Publishers, 1992), p. 258, chpt. 13: 'Ego autem dico, quod omnis operatio causae agentis supra rem aliquam est secundum proportionem materiae recipientis ipsam operationem'. In English: 'Therefore, I say that every operation of the agent of a cause upon a thing is according to the proportion of the material receiving that operation'. I consider the debate surrounding the authorship of the *Speculum* in chapter one of Hendrix, *The Speculum astronomiae and Four Centuries of Readers.*

[27] See Albert the Great's 'De fato', Paul Hossfeld, ed., in *Opera Omnia* (Monasterii Westfalorum: Aschendorff, 1975), pp. 65–78, for an exposition of this understanding of fate.

[28] Albert, 'De quindecim problematibus', p. 36. 'Omni ergo modo ridículo sum est, quod dicunt'.

Proclus, was composed by Aristotle.[29] This point is important not only for
the way if affected Albert's understanding of the transmission of
influences, but also because of the way it would make his system for
understanding the transmission of influences compatible with the systems
Renaissance intellectuals – with their love of Plato – typically held.

Albert did not deal with each question posed to him in such detail. For
Albert this was a settled issue and clear to all except those who display
complete ignorance.[30] Since celestial objects, as corporeal bodies, could
only directly affect the body, not the soul, the will, which is a component
of the intellectual soul, is free to resist corporeal impulses imparted by the
stars. To explain this idea Albert cites the maxim, 'the wise man will
dominate the stars,' a rationale drawn directly from Albumasar's
Introductorium maius (*Greater Introduction*) – though Albert erroneously
attributes this concept to Ptolemy.[31] Albert even goes so far as to argue that
one learned in the influences of the heavens can avert many negative
things, while maximizing positive effects – if one only makes the willed
effort to do so. After all, the impulses relayed through the body affect the
sensible rather than the intelligible soul, which is common to both humans
and animals. This aspect of the soul imparts sensory input capable of

---

[29] Therese Bonin, 'The origin of Diversity in Albertus Magnus', *De Causis et
Processu Universitatis a Prima Causa* (Unpublished dissertation: Notre Dame,
1993), pp. 4, 32–36; de Libera, *Albert le Grand et la Philosophie*, p. 24.

[30] Albert, 'De quindecim problematibus', p. 36. 'Si enim VI liber Primae
Philosophiae legitur, facile patet, qualiter ea quae in inferioribus aguntur,
superiorum subsunt regimini'. In English: "For if book VI of the First Philosophy
is read, it is easily understood, to what degree those things which occur in the
sublunar realm, are to be placed beneath the direction of the supralunar'. For those
who do not understand that which is 'easily understood' from a reading of 'book
VI of the first philosophy [Aristotle's metaphysics]'? 'omnino pateat eorum
ignorantia'.

[31] For Albumasar as the source of this maxim in the West, see Lemay, *Abu
Ma'shar*, pp. 42–48. Medieval intellectuals commonly, though mistakenly, cited
Ptolemy as the source of this concept. G. W. Coopland attempts to trace the
provenance of this maxim in appendix four of his work, *Nicole Oresme and the
Astrologers: A Study of his Livre de Divinacions* (Boston: Harvard University
Press, 1952), pp. 175–77. One should note that Coopland does not give any
indication that it had entered the vocabulary of Latin Christian writers prior to
Albert's use of the saying. Paola Zambelli notes that Albert 'cherished' the
dictum, quoting it repeatedly. See Zambelli's 'Albert le Grand et l'astrologie',
Recherches de théologie ancienne et medieval 49 (1982): pp. 141–58, 146–47.

generating corporeal desires ranging from lust to rage; these 'fancies' or 'desires' regularly motivate people to act without exercising the will making it possible – perhaps even normative – to act according to an impulse of the sensitive appetite.[32] But it is possible to counteract these impulses, especially if one understands their derivation. Such understanding can be acquired by studying the way the heavens affect the human body, thereby allowing the will to function more freely and in greater accord with God's precepts by correcting for celestial influences as needed.[33]

Going a step further, it was this notion that an individual armed with the proper knowledge could manipulate and alter celestial influences that made Albert an advocate of the use of astrological images. These were objects with astrological engravings designed to harness celestial power in such a way as to bring about a wide range of effects, from controlling pests to warding off the evil influences of a malefic planet or affecting the judgment of kings, or even bringing ruin to a city.[34] Nicholas Weill-Parot has argued that the *Speculum astronomiae* introduced the concept of image magic to Latin Christendom, which means Albert the Great if you accept him as the author of this work.[35] Whether or not that is the case, Albert

---

[32] Albert the Great, *Summa Theologiae sive De Mirabili Scientia Dei. Libri I, Pars I, Quaestiones1–50A,* in *Opera Omnia,* edited by Dionysius Siedler, P. A., Wilhelm Kubel, and Heinrich George Vogels, (Monasterii Westfalorum: Aschendorff, 1978), pp. 219–20. Desires 'non sint per essentiam de natura rationis, participant tamen cum ratione.' In English : 'they are not essentially from the nature of reason, but nevertheless they participate with reason.' I explore Albert's understanding of the will and the way in which one may improve the freedom of the will in Scott E. Hendrix, 'Choosing to be Human: Albert the Great on Self Awareness and Celestial Influence,' in *Culture and Cosmos* 12, no. 2, (2008): pp. 23–41. In scholastic terminology the sensitive appetite is the force that provokes an agent to action through the corporeal senses. Properly speaking, the will is associated only with the intellectual appetite.

[33] Albert, 'De fato', pp. 68–69.

[34] Albert, 'De fato', pp. 68–69. Thabit bin Qurra al-Harrani, *De imaginibus,* in Frances Carmody's *The Astronomical Works of Thabit b. Qurra* (Berkeley: University of California Press, 1960), pp. 181, 182, 188.

[35] On images, see Nicolas Weill-Parot, *Les 'images astrologiques' au moyen âge et à la renaissance: spéculations intellectuelles et pratiques magiques (XIIe-XVe siècle)* (Paris: Champion, 2002). I see Albert's authorship of this work as beyond dispute, as I argue in Hendrix, *How Albert the Great's Speculum Astronomiae Was Interpreted and Used by Four Centuries of Readers,* ch. 1. For an alternative view, see Jeremiah Hackett, 'Albert the Great and the Speculum Astronomiae: The State

expressed his confidence in the power of images in his *De mineralibus*.[36] Albert's belief in the power of astrological images enhanced his argument that the study of astrology did not threaten free will, by providing a mechanism whereby one could harness and control celestial influences. How could one argue that the influence of the stars negate free will, if a person is able to control that influence?

The point is, Albert crafted a model of judicial astrology that would allow for predictions that were likely to be true but by no means guaranteed. According to Albert, only when a person is acting in a purely logical fashion can an individual be free of outside influence, including that exerted by celestial bodies.[37] But since people are normally carried along by their impulses astrologers can make 'conjectures' based on celestial observations with a high degree of accuracy in most cases.[38] Furthermore, he created a theological justification in which the study of celestial influences not only did not compromise the free choice of the will, but in fact enhanced it. This view of astrology would prove to be extremely useful, not only to those with a direct connection to Albert such as Thomas Aquinas,[39] but also for a wide range of scholars from Pietro d'Abano (1257–1316) to Giovanni Pontano (1429–1503).[40] However, Albert's defense was not convincing to everyone. In fact, Laura Ackerman Smoller

---

of the Research at the Beginning of the 21st Century,' *A Companion to Albert the Great: Theology, Philosophy, and the Sciences*, edited by Irven Resnick, (Leiden: Brill, 2013): pp. 437–50.

[36] Thorndike, *HMES*, II: pp. 555–62.

[37] Thorndike, *HMES*, II: pp. 234.

[38] Albert the Great, *Questiones. Opera Omnia*, edited by Albert Fries, Wilhelmus Kübel, and Henryk Anzulewicz, (Monasterii Westfalorum: Aschendorff, 1993), p. 59. 'astronomi non dant principia, ex quibus contingit prognosticari aliquid de his quae subsunt libero arbitrio, secundum quod subsunt illi, sed coniecturantur de dispositionibus corporum, quae inclinare et retrahere liberum arbitrium, sicut corpus trahit animam'. In English: 'Astrologers do not give the principles, about those things that are under the power of free will, by which something that is to come to pass may be foretold, but one may make conjectures about these things from the dispositions of bodies, which incline and drag at free will, just as a body drags at the soul'.

[39] Scott E. Hendrix, 'Contemplating the Stars and Comprehending Humanitas: Thomas Aquinas on Astrology and the Essence of what it is to be Human', *History and Culture: Essays on the European Past*, edited by Nicholas C. J. Pappas, (Athens: ATINER, 2012), pp. 11–22.

[40] Hendrix, *Albert the Great's Speculum Astronomiae*, pp. 163–71; p. 209.

has pointed out that the fourteenth century saw the 'most thoroughgoing attack on judicial astrology in the West since Augustine'.[41] Nicole Oresme (1320–1382) and his younger contemporary, Henry of Langenstein (d. 1397), were the primary architects of this challenge to astrology.[42]

However, we should be cautious about seeing the opposition – no matter how vigorous it might have been – of writers such as Nicole of Oresme or Henry of Langenstein as normative for the fourteenth century. During those years, several events would occur that would leave Europe's intellectual elites scrambling to find explanations. I have already mentioned the Black Death above, but two other events of special importance to our current study are the Avignon Papacy (1309–1377) and the Great Schism (1378–1417).[43] Jointly these events would do tremendous harm to the prestige of the Church while promoting movements dangerous to traditional Catholicism. Particularly problematic were the apocalyptic visionaries who arose on the continent, men such as Vincent Ferrer (1350–1419) whose 1398 vision of the Last Judgment led him to a twenty-year ministry that many found quite alarming.[44]

The excesses of those preaching apocalypticism made astrology quite attractive to some. For example, the theologian and cardinal Peter d'Ailly (1350–1420) initially opposed the study of astrology, fearing that it might lead Christians to ignore theology.[45] However, as he aged, d'Ailly's

---

[41] Laura Ackerman Smoller, *History, Prophecy, and the Stars* (Princeton: Princeton University Press, 1994), p. 32.

[42] G. W. Coopland, *Nicole Oresme and the Astrologers: a study of his De Divinacions* (Liverpool: Liverpool University Press 1952); Ronal Brashear, 'Henry of Langenstein,' *Biographical Encyclopedia of Astronomers* (New York: Springer, 2014): pp. 913–14; Nicolas H. Steneck, *Science and Creation in the Middle Ages: Henry of Langenstein (d. 1397) on Genesis* (Notre Dame: University of Notre Dame Press, 1976).

[43] For a brief overview of these events and their impacts, see Christopher M. Bellitto, *Renewing Christianity: A History of Church Reform from Day One to Vatican II* (Mahwah: Paulist Press, 2001), pp. 106–12; Walter J. Woods, *Walking with Faith: New Perspectives on the Sources and Shaping of Catholic Moral Life* (Collegeville: The Liturgical Press, 2010), pp. 333–42.

[44] Bernard McGinn, *Visions of the End: Apocalyptic Traditions in the Middle Ages* (New York: Columbia University Press, 1998), pp. 253–58.

[45] Smoller, *History, Prophecy, and the Stars*, 47, quoting Pierre's *Tractatus utilis super Boecii*, 160r: 'Nolunt ergo sancti quod relicta morali philosophia vel sciencia que pertinet ad anime salute astrologia studeatur'. In English: 'Therefore, holy men should not wish that moral philosophy or science should be abandoned, which pertains to the salvation of the soul, so that astrology might be studied'.

position matured. In the opening years of the fourteenth century, he summarized his position in the *De Concordia theologiae et astronomiae* (*On the Concord of Theology and Astronomy*): 'I do not reject the truth of astrology but the vanity within certain of the astrologers'.[46] For d'Ailly, astrology represented a rational means of understanding the calamities of the world that would prove quite soothing to the French cardinal's attempts to come to terms with these events, while simultaneously providing an alternative to the dangerously unverifiable (and therefore uncontrollable) prophecies of doom that multiplied in the fourteenth century.[47] In contrast to these prophecies, astrology was replicable, verifiable, and controllable. In a letter to Jean Gerson, d'Ailly explained as much while appealing to none other than Albert the Great for support.[48]

Is it surprising that d'Ailly found the logic of astrology preferable to the socially caustic rhetoric of the Antichrist that prophets such as Ferrer spread? And for those who did not see astrology in the same light, Pierre d'Ailly wrote the *Vigintiloquium de concordantia astronomice veritatis* (*Twenty Statement about the Concord of True* Astrology) while at Cologne in 1414 in order to establish the validity of astrological divination in opposition to unverifiable prophetic visions. Stating that 'it is necessary to harmonize true astrology with sacred theology,' d'Ailly then proceeds to consider twenty points of contention that might arise between conservative theologians and adherents of astrology.[49]

---

[46]Pierre, *Tractatus utilis super Boecii*, f. 140r. 'Ego autem non astronomie veritatem sed quorumdam astronomorum vanitatem secum reprobo'.

[47] Smoller, *History, Prophecy, and the Stars*, pp. 92–95.

[48] Pierre d'Ailly, *Apologia defensiva astronomiae ad magistrum Johannem cancellerium parisiensem* (Louvain: J. de Paderborn, 1483), f. 143v: 'Concordemus denique cum Alberto magno doctore sancti Thomae in illo praecipue tractatu suo qui Speculum dicitur, ubi hanc materiam plene utiliterque pertractat'. In English: 'I agree, then, with Albert, the great learned professor of sainted Thomas, especially in his tract which is called the Speculum, where he deals with this material most fully and usefully'.

[49] Pierre, *Apologia defensiva astronomiae*, f. 2r. This is part of d'Ailly's opening statment, one of his 'viginiti continens verba feliciter incipit' that gave the work its name. In its entirety, d'Ailly states: 'Primum: secundum philosophum omne verum omni vero consonat; necesse est veram astronomie scientiam sacre theologie concordare'. In English: 'Point one: according to all philosophy which is consonant with truth, it is necessary that the true science of astrology agree with sacred theology'. After listing his twenty propositions, he then goes on to explicate each one fully in proper scholastic fashion, beginning on 3r.

Seeking to get to the root of the opposing camp's position, he details errors found in illicit works of astrology, before stating that 'the aforementioned errors had been reproved not only by sacred theologians, but also by true astronomers'.[50] All of these errors are ultimately traceable to those 'who call fate the force of the position of the stars and of the constellations in which all things occur in these inferior parts by necessity'.[51] However, there is no such thing as 'fate', in the sense of foreordained unalterable outcomes, and in relation to this 'doctors of theology have proven sufficiently, with whom Ptolemy, the most skilled of the astronomers, does not disagree, when he says that the prudent man rules the stars'.[52] This, then, represents the concord between astronomy and theology that gave his work its title: according to d'Ailly, both of these *scientiae* recognize that heavenly bodies may dispose and incline individuals toward certain types of behavior, but cannot predicate actions from necessity. In other words, he had taken up the arguments Albert the Great had developed in the prior century.

The most commonly accepted understanding of astrology by the fifteenth century was that it was seen as consistent with Catholic theology and the doctrine of free will; and that the study of astrology could allow greater freedom of the will. Therefore, given the privileged position of free will and human potential in the Renaissance, it is natural to find a champion of astrology in one of the most important philosophers of the humanist movement: Marsilio Ficino (1433–1499).[53] Although he followed

---

[50] Pierre d'Ailly, *Vigintiloquium de concordantia astronomicae veritatis cum theologia* (Venice: Erhardus Ratdolt, 1490), f. 3r–3v. Peter classifies these errors within three categories: deterministic beliefs, the mingling of superstition in otherwise sound works, and those theories of astrology that compromise free will by allowing too much power to divine and supernatural forces; Pierre, *Vigintiloquium,* f. 3r. 'Prefati errores non solum a sacris theologis: sed etiam a veris astronomis fuerunt reprobati'.

[51] Pierre, *Vigintiloquium,* f. 3r. 'Vocant fatum vim positionis siderum et constellationum in quam omnia in his inferioribus necessario eveniunt'.

[52] Pierre, *Vigintiloquium,* f. 3r. 'Doctores theologi sufficienter probaverunt [thatis: qua fatum nihil fit] a quibus non discordat peritissimus astronomorum Ptolomeus: ubi ait quod vir prudens dominatur astris'.

[53] Paul Oskar Kristeller, *Eight Philosophers of the Italian Renaissance* (Stanford: Stanford University Press, 1964), pp. 37–53; Paul Oskar Kristeller, *The Philosophy of Marsilio Ficino* (New York: Columbia University Press, 1943); Graziella Federici Vescovini, 'The Theological Debate,' *A Companion to Astrology in the Renaissance*, edited by Brendan Dooley, (Leiden: Brill, 2014): pp. 99–140.

in his father's footstep and trained as a physician, he was equally inspired by the fifteenth-century revival of Greek linguistic study and Platonic philosophy and applied himself to a rigorous study of both. As outlined in his most important original work, the *Theologia Platonica de immortalitate animae* (*Platonic Theology about the Immortality of the Soul*), he accepted a Neoplatonic model of the celestial hierarchy reaching down from God to humankind, giving him a deep interest in astrology throughout his life.[54] At times Ficino's contemporaries found his advocacy of judicial astrology to be too extreme, as when he was called to defend his writings before Pope Innocent VIII in 1489,[55] but as we will see, such concerns were based on misunderstanding of Ficino's position regarding astrology.

Ficino warns his readers that astrological determinations can be difficult because the distances involved between terrestrial creatures and the heavens can lead to tremendous errors in understanding the influences imparted.[56] Furthermore, proximate influences, from ancestry to diet, are more powerful than that imparted by the distant stars.[57] He also cautions that we should never forget the overpowering nature of divine influence, stating that his own career is the result of a yearning toward knowledge implanted by God rather than the product of interacting mediating substances, but he also states that astrology has provided him with a means to understand his own misfortunes, and that he has used it to advise friends

---

[54] Arthur Field, 'The Platonic Academy of Florence', *Marsilio Ficino: His Theology, His Philosophy, His Legacy*, edited by Michael J. B. Allen, Valery Rees, and Martin Davies, (Leiden: Brill, 2002), pp. 359–73; Thomas Moore, *The Planets Within: The Astrological Psychology of Marsilio Ficino* (Great Barrington: Lindisfarne Books, 1989); M. Bullard, 'The Inward Zodiac: A Development in Ficino's Thought on Astrology', *Renaissance Quarterly* 42 (1990): pp. 687–708.

[55] Jonathan Arnold, *The Great Humanists: European Thought on the Eve of the Reformation* (London: I. B. Tauris, 2011), p. 66.

[56] Don Cameron Allen's discussion of Ficino's views is commendable, though he errs in his final interpretation of Ficino's position. See Don Cameron Allen, *The Star-Cross Renaissance: The Quarrel about Astrology and its Influence in England* (Durham: Duke University Press, 1941), pp. 11–19.

[57] Marsilio Ficino, 'De vita coelitus comparanda', book three of his *De vita libri tres*, edited by Martin Plessner, (New York: George Olms Verlag, 1978), chs. 1–8, 15. This volume is a reproduction of a 1498-printed edition of the work. It is not paginated and the chapters are quite short. For the sake of convenience and clarity, I cite by book and chapter, rather than by page.

and forewarn patrons about impending difficulties.[58] All of this suggests a traditional view of celestial influence and astrology: the heavens impart impulses that interact with proximate causes in the body to move us toward an action, but we can always engage our free will to overcome these impulses.[59] Ficino sounds much like Albert when discussing celestial influence and the ability to manipulate it, and small wonder, because the Renaissance Italian regularly drew directly on his German predecessor.[60]

What we see in Ficino's work is a consistent belief in the predictive powers of astrology and the usefulness of the *scientia* in many spheres of human activity. When he warns his readers about the use of predictive astrology, it is not a rejection, but rather exactly what he says it is: a cautionary note about the complex web of influences that affect terrestrial creatures, including but not limited to powerful proximate causes that make accurate predictions difficult, though not impossible. Such warnings were common enough in the Middle Ages and later.[61] Therefore, when one

---

[58] Ficino, 'De vita longa', in *De vita libri tres*, ch. 3.

[59] Ernst Cassirer, *Individuum und Kosmos in der Philosophie der Renaissance* (Leipzig: Teubner, 1927), pp. 120–21.

[60] Ficino, 'De vita coelitus comparanda', in *De vita libri tres,* chapter 12; Yates, pp. 73–74. Ficino refers to Albert the Great in his *De vita libri tres*, ch. 19, as 'astrologiae partiter atque theologiae professor [...] a licitis discernere se inquit illicita'. In English: 'A professor equally of astrology and theology [. . .] who says he is able to discern the licit from the illicit [in astrology]'. On the question of free will, Ficino writes: 'Albertus quoque Magnus inquit in speculum, non enim libertas arbitarii ex electione horae laudabilis coercetur, sed potius in magnarum rerum inceptionibus electiones horae contemnare est arbitrii praecipitatio non libertas'. In English: 'Albert the Great also says in the Speculum, that the freedom of the will is not coerced by the election of a favorable hour, but rather, to condemn elections in the beginnings of great things is to throw away judgment, not freedom'. Ficino, 'De vita coelitus comparanda', in *De vita libri tres*, ch. 12.

[61] The intricacies and difficulties of astrology were a common trope. Writing in 1391, the anonymous compiler of Bodley 581, produced for King Richard II (1377–1399) of England, states that 'the science of astronomy is both of great difficulty and is time-consuming to learn, for which the present life is scarcely adequate'. Quoted in Hilary Carey, *Courting Disaster: Astrology at the English Court and University in the Later Middle Ages* (New York: St. Martin's Press, 1992), p. 103. Albert had earlier acknowledged the difficulties involved in his 'De fato', written at Anagni in 1256. According to Albert, 'in caelesti circulo quoad nos infinita consideranda sunt, sicut stellae in numero et specie et virtutibus et situs earum in circulo declivi et extra ipsum et distantiae et coniunctiones et quantitas anguli, sub quo incidit radius, et pars fortunae et gradus lucidi et umbrosi

reads Ficino's comments within their proper context, informed by an understanding of pre-modern astrological beliefs, we can see that rather than a rejection of judicial astrology based upon proto-modern skepticism, we have a thoroughly traditional caveat about the complexities involved in prognosticative efforts.

---

in puteis et in turribus existentes et huiusmodi infinita quoad nos'. In English: 'in the circle of heaven, there are things we should consider infinite, such as the number of the stars as well as their species, powers, and their declination in their circuit, and beyond that, their distances, conjunctions, and the quantities of their angles, beneath which a ray falls, as well as the part of fortune and the grade of the light and shadow standing in wells and on towers and infinite things of this sort', 'De fato', 72. However, 'multa et quoad nos infinita consideranda essent, sed considerantur paucissima, quibus oboediunt alia, et ex illis pronosticabilis habetur coniecturatio. Propter hoc dicit Ptolemaeus, quod elector non nisi probabiliter et communiter iudicare debet.' In English: 'There are many things regarding which should be considered infinite, but very few are taken into account, regarding which they [astrologers] take into account some [of the many things that can be considered infinite] and a conjecture is made from those prognostications. For this reason, Ptolemy states that a prognosticator should not make a judgment except in probably and general terms', 'De fato', 73.

Poetry Creation as Space of Union between Natural and Supernatural: A Reading of *The House of Fame*

Gerardina Antelmi

Abstract: Cosmos, planets and astrology occupy a significant place in medieval dream poems, which have been analysed as models of cosmological dream allegories, supernatural voyages, and intellectual journeys exploring the universe. The dreamer often soars above the Earth acquiring a privileged view of the Earth and the Cosmos. However, it seems to me that similarities between the ascending structure of some medieval literary works and the mystical ascent towards the mystical marriage with the Divine has been overlooked.

In this paper I will examine Chaucer's *The House of Fame*. After reviewing the stages recognized by medieval mystics to seek mystical union with the Divine, I will identify specific linguistic markers employed by Chaucer, taken from the language of mysticism, which serve as signposts for the dreamer's ascent to the sky. I will then illustrate how the dream mirrors the mystical journey, how the Introduction of the poem can be interpreted as a microcosm and the three books of the poem as macrocosm.

I will argue, finally, that the ascent of the dreamer-poet to the sky to the House of Fame represents the poet bridging the material and the supernatural, Earth and Heaven – and that the dreamer-poet's possible 'stellification' symbolizes that poetry is a sacred space in which this union becomes possible.

In the past, in order to travel into the Cosmos two devices were available: flying animals and dreams.[1] The dream poem the *House of Fame* by Chaucer offers both channels: the dream is a vehicle in which to depart from earth and to reach the beyond, and the golden eagle, the animal with which the dreamer flies. In this study I aim to demonstrate that in medieval dream visions the dreaming poet bridges the material and the supernatural, and that poetry creation is represented as occurring during an enhanced

---

[1] Dean Swinford, *Through the Daemon's Gate. Kepler's Somnium, Medieval Dream Narratives, and the Polysemy of Allegorical Motifs* (New York and London: Routledge, 2006), p. 46.

Gerardina Antelmi, 'Poetry Creation as Space of Union between Natural and Supernatural: A Reading of *The House of Fame*', *The Marriage of Heaven and Earth*, a special issue of *Culture and Cosmos*, Vol. 20, nos. 1 and 2, 2016, pp. 147–66.
www.CultureAndCosmos.org

state of awareness, a sacred space where the union between natural and supernatural is possible. On investigation, Chaucer's dream poem reveals an inner structure that resonates with the hierarchy of stages leading to ecstasy as envisaged by the mystics.

It is no easy task to examine concisely two fields as rich and complex as medieval dream visions and the mystics' contemplative journey, and the result may not quite satisfy the specialists of either field. Although the topic would benefit from an exhaustive cross-disciplinary exploration of mysticism and dreams, literary and non-literary, including ethnological studies, here I will limit the scope of the analysis to the writings of the Middle English mystics – Richard Rolle (1290/1300–1349), Walter Hilton (1340/45–1396), Julian of Norwich (1342–1416), the anonymous author of The *Cloud of Unknowing* (ca. 1375), and Margery Kempe (ca. 1373–1440) – the *House of Fame* (*HF*), and to specific ethnological studies in passing. I will first illustrate the mystical journey to contemplation, as outlined by Middle English mystics. The relevant terminology, crucial figurative imagery and symbols appearing in the *HF* will be signposted and paralleled with those pertaining to the mystical ascent. Reading *HF* from such a perspective does not imply or signify that Chaucer is to be deemed a mystic. He was not a mystic; neither the *HF* nor his main works are mystical. Nonetheless, the 'cultural and social intertext' constituted by devotional writings might have influenced Chaucer, be it unintentionally.[2] The medieval cultural texture was imbued with spirituality to such an extent that an audience today can barely separate religious aspects from the whole medieval life experience, given the 'all-pervasiveness of religion' at that time.[3]

The abundance of dream visions in the period 1350–1400 has been acknowledged and examined through a number of lenses.[4] These works,

---

[2] Robert Boenig, *Chaucer and the Mystics: The Canterbury Tales and the Genre of Devotional Prose* (Lewisburg, PA: Bucknell University Press, 1995), pp. 9–11.
[3] Helen Cooper, 'Introduction', in *Chaucer and Religion*, ed. Helen Phillips (Woodbridge, Suffolk, and Rochester, NY: D. S. Brewer, 2010), p. xv.
[4] Walter Clyde Curry, *Chaucer and the Medieval Sciences* (1926; repr. London: Allen & Unwin, 1960); Francis X. Newman, 'Somnium: Medieval Theories of Dreaming and the Form of Vision Poetry' (PhD diss., Princeton University, 1962); Constance B. Hieatt, *The Realism of Dream Visions. The Poetic Exploitation of the Dream-Experience in Chaucer and his Contemporaries* (The Hague-Paris: Mouton & Co., 1967); Anthony C. Spearing, *Medieval Dream Poetry* (Cambridge: Cambridge University Press, 1976); Kathryn Lynch, *The High Medieval Dream*

which narrate 'dreams dreamt either by the narrator of the poem (...) or by one of his characters', have been analysed according to the dream lore that the Middle Ages inherited from the classical world, in particular from the pagan philosopher Macrobius, the authority in dream-lore, whose classification will be illustrated shortly.[5] After the symbolic interpretation by the exegetical school, recent criticism has emphasized the psychological aspects of the dreamer-narrator; considered medieval dream visions as secular works; identified poetry composition as their main topic; and recently restated the religious aspects impacting on Chaucer's works.[6] Helen Cooper rightly asserts that religious convictions have not been treated seriously by scholarship, and further observes that the 'mystical experience was something more, or other, than neurosis'.[7] Studies have associated poetry with meditation, but, to my knowledge, dream visions, employed extensively by Chaucer, have yet to be paralleled to the mystics' experiences.[8] Investigating Chaucer's dream visions from the perspective of medieval mysticism will shed new light on this literary genre.

Vision: Poetry, Philosophy, and Literary Form (Stanford, CA: Stanford University Press, 1988); J. Stephen Russell, *The English Dream Vision. Anatomy of a Form* (Columbus, OH: Ohio State University Press, 1988); Steven F. Kruger, *Dreaming in the Middle Ages* (Cambridge: Cambridge University Press, 1992); Peter Brown, ed., *Reading Dreams: the Interpretation of Dreams from Chaucer to Shakespeare* (Oxford: Oxford University Press, 1999).

[5] Spearing, *Medieval Dream Poetry*, p. 1. For an outline of medieval dream classifications see also Curry, *Chaucer and the Medieval Sciences*; Kruger, *Dreaming in the Middle Ages*.

[6] For the interpretation of the exegetical school, see Bernard F. Huppé and Durant W. Robertson Jr, *Fruyt and Chaf: Studies in Chaucer's Allegories* (Princeton, NJ: Princeton University Press, 1963); B. G. Koonce, *Chaucer and the Tradition of Fame: Symbolism in 'The House of Fame'* (Princeton, NJ: Princeton University Press, 1966); as to the psychological perspective, Spearing, *Medieval Dream Poetry*; Paul Piehler, *The Visionary Landscape: A Study in Medieval Allegory* (London: Edward Arnold, 1971); for the *HF* and writing, see Robert M. Jordan, *Chaucer's Poetics and the Modern Reader* (Berkeley, CA: University of California Press, 1987), pp. 22–50; for a discussion on religion, see Helen Phillips, ed., *Chaucer and Religion* (Woodbridge, Suffolk, and Rochester, NY: D. S. Brewer, 2010); Charles Muscatine, 'Chaucer's Religion and the Chaucer Religion', in *Chaucer Traditions. Studies in Honour of Derek Brewer*, ed. Ruth Morse and Barry A. Windeatt (Cambridge: Cambridge University Press, 1990), pp. 249–62.

[7] Cooper, 'Introduction', p. xi.

[8] Boenig examines the impact of mysticism on the *Canterbury Tales* in *Chaucer and the Mystics*. For a study on the relation between meditation and poetry of a

The Mystical Ascent

In different epochs and cultures, humankind has attempted to reach an enhanced level of consciousness. Mystics share the belief that the ultimate reality is not of the earthly world – a world from which they wish to detach – but belongs, rather, in an unseen world.[9] Their main purpose is to achieve this reality. Nonetheless, the individual's will to successfully reach divinity – which is an act of love towards divinity – is not sufficient, as the highest degrees of contemplation represent a gift yielded by divine grace.[10] Will, love, and grace represent one of the many triads characterising the systemization of the ascent. Whilst visions could occur spontaneously, habitual disciplined exercises to induce the contemplative state were also practised. Within the mystical ascent to contemplation – be they mystics reporting of the other world or devotional treatises addressed to anchorites – several stages and relative techniques are evident. The first step is to tame the body and mind through sleep deprivation, fasting and self-injurious behaviour, upon which the medieval instructors generally recommend moderation.[11] The purpose of controlling the mind is to

---

later period as well as an outline of medieval meditative practices, see Louis Lohr Martz, *The Poetry of Meditation: a Study in English Religious Literature of the Seventeenth Century* (New Haven, CT: Yale University Press, 1955). On the employment of 'meditatio' in late-medieval religious writings, see Thomas H. Bestul, 'Chaucer's Parson's Tale and the Late-Medieval Tradition of Religious Meditation', *Speculum* 64, no. 3 (1989): pp. 600–619. Barbara Newman mentions the possibility that devotional visualization can be combined with poetry composition, in 'What Did It Mean to Say "I Saw"? The Clash between Theory and Practice in Medieval Visionary Culture', *Speculum* 80, no. 1 (2005): pp. 1–43, p. 17. For parallels with monastic memory practices in works written until the thirteenth century, see Mary Carruthers, *The Craft of Thought. Meditation, Rhetoric, and the Making of Images, 400-1200* (Cambridge: Cambridge University Press, 1998).

[9] Jerome Kroll and Bernard Bachrach, *Mystic Mind: the Psychology of Medieval Mystics and Ascetics* (Abingdon, NY: Routledge, 2005), pp. 48–52. See also Evelyn Underhill, *Mysticism: a Study in the Nature and Development of Man's Spiritual Consciousness* (London: Methuen, 1960), p. 65.

[10] Patrick J. Gallacher ed., *The Cloud of Unknowing* (Kalamazoo, MI: Medieval Institute Publications, Western Michigan University, 1997), Chapter 8; at http://d.lib.rochester.edu/teams/text/gallacher-cloud-of-unknowing   [accessed October 2014].

[11] Walter Hilton, *The Scale of Perfection*, ed. Thomas H. Bestul (Kalamazoo, MI: Medieval Institute Publications, Western Michigan University, 2000), Book I,

interrupt the course of trivial thoughts which flow incessantly there within. Richard Rolle compares the inner state of peace and silent contemplation, when thoughts cease in the mind, to a 'haliday' or a 'saterdai in hir soul'.[12] To prevent the onset of sleep, and to prolong the waking, mystics turn to prayer, which also fosters transition to another state of consciousness. Subsequent to the physical taming, the basic gradation towards contemplation is composed of three stages.[13] The Middle English mystics suggest 'Lesson, Meditacion, and Oryson. Or elles (…) Redyng, Thinkyng, and Preiing': reading the Scriptures, meditating, and praying.[14] Prayer, practised throughout the stages, is of three types: vocal prayer; prayer derived directly from the heart yielded by God; and in the higher stages of contemplation, if prayer consists of words, these should be very short or of only one syllable. However, the highest form of prayer is speechless.[15] Similarly, as the highest prayer is emptied of words, the most elevated visionary experiences lack images.[16] According to Kolve, 'the persistence of triads must be noted – they honour the Trinity, and constitute the discovery, by medieval man, of the image of the Creator in himself'.[17] Hugh

---

Chapters 2, 19, 22, at http://d.lib.rochester.edu/teams/publication/bestul-hilton-scale-of-perfection [accessed November 2014]; Richard Rolle, 'Form of Living', in *English Writings of Richard Rolle*, ed. Hope E. Allen (Oxford: Clarendon Press, 1931), ch. 5, p. 96. For an outline of the preparatory techniques to meditation, see also Martz, *Poetry of Meditation,* pp. 25–39.

[12] Rolle, *The Mendynge of Lyfe*, in *English Mystics of the Middle Ages*, ed. Barry A. Windeatt (Cambridge: Cambridge University Press, 1994), p. 22.

[13] Martz, Poetry *of Meditation*, p. 34. Elizabeth Buckmaster parallels the tripartite structure both of the poem and of meditation in 'Meditation and Memory in Chaucer's *House of Fame'*, *Modern Language Studies* 16, no. 3 (1986): pp. 279–87, p. 279.

[14] Gallacher, *Cloud*, ch. 35, ll. 1312–13; see also Hilton, *Scale*, Book I, ch. 15, ll. 333–34; Rolle, *The Mendynge of Lyfe*, pp. 18–19.

[15] Gallacher, *Cloud*, ch. 37, ll. 1367–69; ch. 47, ll. 1644–47; Julian of Norwich, *The Shewings of Julian of Norwich*, ed. Georgia Ronan Crampton (Kalamazoo, MI: Medieval Institute Publications, Western Michigan University, 1994), ch. 13, ll. 500–501, at http://d.lib.rochester.edu/teams/publication/crampton-shewings-of-julian-norwich [accessed November 2014].

[16] Barbara Newman, 'What Did It Mean to Say "I Saw"?', p. 12.

[17] V. A. Kolve, 'Chaucer and the Visual Arts', in *Writers and their Background: Geffrey Chaucer*, ed. Derek Brewer (Athens, OH: Ohio University Press, 1975), p. 301, as quoted in Buckmaster, 'Meditation and Memory', p. 280. For speculations on the tripartite structure in the *HF*, see Piero Boitani, 'Old Books brought to life

(1096–1141) and Richard of St Victor (d. 1173) – who deemed the contemplative state could be reached through dreams – classify the journey to the ultimate reality through *cogitatio, meditatio, contemplatio*.[18] This triad carries relevant overtones for the following analysis of the terminology employed in the introductory lines of the *HF*. Although most of the hierarchical systemization of the path to ecstasy is tripartite, some devotional writings put forward an ecstatic journey detailed in a series of rungs. In *Scala claustralium* Guigo II suggests the contemplative should climb four rungs up the ladder to contemplation: reading (*lectio*), reflection (*meditatio*), prayer (*oratio*), and contemplative trance (*contemplatio*), whereas Bonaventure identifies seven degrees in the ascent to God, wherein the real mystical ascesis can only be achieved through the seventh stage: 'mysticum' signifying 'secret', known only by those who desire to attain it.[19]

The journey to the higher stages of contemplation is interspersed with a variety of symbols, including the door (or gate), and the flight (or else wings, or birds).[20] The 'streite gate', marking the passage from meditation

---

in dreams: the *Book of the Duchess*, the *House of Fame*, the *Parliament of Fowls'*, in *The Cambridge Companion to Chaucer*, 2nd ed., ed. Piero Boitani and Jill Mann (Cambridge: Cambridge University Press, 2003), pp. 58–77, pp. 73–74.

[18] Hugh of Saint Victor, *De modo dicendi et meditandi*, *Migne Patrologia Latina (PL)*, 176, col. 879, at http://www.documentacatholicaomnia.eu/04z/z_1096-1141__Hugo_De_S_Victore__De_Modo_Dicendi_Et_Meditandi_Libellus__MLT .pdf.html [accessed November 2014]; Richard of Saint Victor, *Benjamin major*, 1, 3, *PL*, 196, col. 66, at http://www.documentacatholicaomnia.eu/04z/z_1162-1173__Richardus_S_Victoris_Prior__De_Gratia_Contemplationis_Libri_Quinque _Dicti_Benjamin_Major__MLT.pdf.html [accessed October 2014]. As to the diffusion of Richard of Saint Victor's works in medieval England, see Boenig, *Chaucer and the Mystics*, p. 23.

[19] Newman, 'What did it mean to say "I saw"?', p. 18. St Bonaventure, *The Soul's Journey into God*, trans. Ewert Cousins (New York, NY: Paulist Press, 1978), p. 113; *Itinerarium mentis in Deum*, http://www.franciscanos.net/document/itinerl.htm [accessed November 2014].

[20] Hildegard of Bingen identifies four means by which the soul can fly, *Book of Divine Works*, ed. Matthew Fox (Santa Fe, NM: Bear & Company, 1987), p 120; for the employment of flying and of the eagle in Mechtild of Magdeburg, and Hadewijch of Brabant, see Elizabeth A. Petroff, *Medieval Women's Visionary Literature* (New York, NY: Oxford University Press, 1986), respectively pp. 217–18, 197–98.

to contemplation, is 'narw, and fewe men fynden it'.[21] In the context of a temple the door represents the threshold between the profane and sacred world.[22] The temple usually has an opening at the top which permits contact with the supernatural world, and has the function of a door through which the gods can descend to the earth and humans can symbolically ascend to heaven.[23] The images symbolising the soul ascending towards the Absolute are often bound with images related to flying. Indeed, Richard of St Victor compares contemplation to a flight of birds.[24] In poetry, flight with an eagle is associated with acquisition of knowledge: whilst asleep Dante dreams of an eagle before being lifted up to the next stage in Purgatory by St Lucy.[25] This episode is echoed by the dreamer's flight in the *HF*. The highest stage – encountering the divinity – is shrouded by a cloud, as *The Cloud of Unknowing* recalls, entering which is terrifying. The sacred has always inspired awe and dread.[26] Moreover, the sacred, the mystics' experience of their encounter with the divinity, is ineffable and cannot be expressed through human words.[27] Hard though it may be, the contemplative state is depicted by mystics as 'the perfect love of God' and essentially as a 'joyinge of the soule', a 'gladnesse unspecable', a 'joy that (…) passyth al that herte may willen and soule may desire'.[28]

---

[21] Matthew 7:13-14, quoted in Hilton, *Scale*, Book I, ch. 42, ll. 1130–32. Rolle states that through revelation the 'dore of hevene' is opened, *The Mendynge of Life*, p. 20.

[22] Arnold van Gennep, *Riti di passaggio* (Torino: Boringhieri, 1981), p. 18.

[23] Mircea Eliade, *Il sacro e il profano* (Torino: Bollati Boringhieri, 1995), p. 22.

[24] Richard of St Victor, *Benjamin major*, I, 5.

[25] Dante Alighieri, *Purgatorio*, IX, 17–21, available at http://www.bibliotecaitaliana.it/indice/visualizza_testo_html/bibit000019 [accessed November 2014].

[26] See Julian of Norwich, *Shewings*, ch. 8.

[27] Julian, *Shewings*, ch. 26, ll. 924–29. Rolle declares that 'þe wonderfull joy of þe kyngdom of heven, thyrd es mare þan tong may tell, or hert mai thynk, or egh may se, or ere may here', 'Form of Living', ch. 6. See also Margery Kempe, *The Book of Margery Kempe*, Lynn Staley ed., (Kalamazoo, MI: Medieval Institute Publications, Western Michigan University, 1996) ch. 17, ll. 900–902, at http://d.lib.rochester.edu/teams/publication/staley-the-book-of-margery-kempe [accessed November 2014]

[28] Respectively Hilton, *Scale*, Book II, ch. 21, ll. 1129–31; Rolle, *The Mendynge of Lyfe*, p. 19; Hilton, *Scale*, Book II, ch. 36, ll. 2539; Julian, *Shewings*, ch. 26, ll. 926–27; see also Margery Kempe, *The Book of Margery Kempe*, ed. Lynn Staley (Kalamazoo, MI: Medieval Institute Publications, 1996), I, ch. 28, available at

Dream Visions

The following part explores how the model of the mystical transition between states of consciousness applies to Chaucer's *HF*, and how within this context the stages from ordinary awareness to the more intense consciousness are signalled by means of symbols shared by both mystics and poets. The dream vision the *HF* was composed around 1380, prior to the *Canterbury Tales*. Divided into three books, it recounts a dream of the narrator, who calls himself Geffrey.[29] In summary, after the Proem wherein the poet refers to medieval dream-lore, Book I is set in a glass temple of Venus, where Geffrey recounts the story of Dido and Aeneas, which is portrayed on the walls. At the conclusion of the narration the dreamer steps through a gate out of the temple into a desert. At the beginning of Book II: the narrator announces that the audience is going to hear such a marvellous dream that not even the famous dreamers of Antiquity (Scipio, Nabuchodonosor, Pharaoh) had ever dreamed of its like. In the dream he is lifted up by the Eagle sent by Jove to carry him to the House of Fame. As the dreamer is lifted up, he faints. When he regains consciousness, the Eagle explains that, though Geffrey has written about love, his experience only comes from books not from experience, so they will go to the House of Fame, a place equidistant from heaven, sea and sky, in search of tidings. During their flight the Eagle speaks of the nature of sound. In Book III they arrive in the realm of Fame where the dreamer is deposited and witnesses how Fame grants or withdraws fame to men by caprice. After leaving this house, he sees a wicker house, sixty miles long, spinning in the air full of 'tydynges'. Given the speed of its rotation, the dreamer needs the Eagle's assistance to enter. Inside motion seems to halt. The poem ends abruptly, apparently incomplete, on the appearance of a Man of Great Authority who seems to be on the verge of delivering news. However, neither the man of authority makes an utterance, nor does the dreamer wake up.

The complexity of the poem, its cryptic nature, the presence of unusual elements for a dream vision (the winter setting; the desert; the wish that

---

http://d.lib.rochester.edu/teams/publication/staley-the-book-of-margery-kempe [accessed November 2014].

[29] To be noted that the final title was *Book of Fame*, see Stephen Knight, 'Classicizing Christianity in Chaucer's Dream Poems: the *Book of the Duchess, Book of Fame* and *Parliament of Fowls*', in *Chaucer and Religion*, ed. Helen Phillips (Woodbridge, Suffolk, and Rochester, NY: D. S. Brewer, 2010), pp. 143–55, p. 143; Jack A. W. Bennett, *Chaucer's Book of Fame* (Oxford: Clarendon Press, 1968), p. ix.

every dream can come to good, as opposed to the belief in dreams) and the supposed incompleteness have all inspired a variety of debates, leaving it open to multiple interpretations.[30] Recent criticism has focused mainly on the comic aspects, including the ironic self-portrait of the poet and the 'garrulous' eagle, and although the *HF* has been viewed as a journey to another world, this acknowledgment is merely peripheral.[31] In contrast, Stephen Knight argues that, though the poem has been considered secular, the religious 'is a recurrent motif of this poem'.[32] The employment of undoubtedly comic features and the presence of terms and symbolic imagery related to mysticism need not be mutually exclusive, as Boenig demonstrated.[33] Here I will bring to the fore elements shared by poetry and mysticism through a detailed examination of terms employed in the introductory section and of symbols used at salient moments in the poem. The first fifty-eight lines, often defined as a review of the different types of dreams and as the illustration of their causes, have also on occasions been deplored.[34] However, an attentive reading will reveal that the section refers

---

[30] For a discussion of possible interpretations, see Steven Kruger, 'Dreaming', in *A Concise Compendium to Chaucer*, ed. Corinne Saunders (Oxford: Blackwell, 2006), pp. 71–89; for scepticism and irony as unifying theme, see Sheila Delany *Chaucer's House of Fame: the Poetics of Skeptical Fideism* (Chicago and London: University of Chicago Press, 1972); for a discussion concerning the conclusion, see Donald Fry, 'The Ending of the *House of Fame*', in *Chaucer at Albany*, ed. Rossell H. Robbins (New York: Franklin & Co., 1975); for parallels with memory and post-modern technologies, see Ruth Evans, 'Chaucer in Cyberspace: Medieval Technologies of Memory and the *House of Fame*', *Studies in the Age of Chaucer* 23, no. 1 (2001): pp. 43–69.

[31] For mentions of the journey to the other world, see Susanna Fein, 'Other Thought-worlds', in *A Companion to Chaucer*, ed. Peter Brown (Oxford and Malden, MA: Blackwell Publishers, 2002), pp. 332–48, p. 332; Kruger, 'Dreaming', p. 85; for speculations on the comic aspects, see Fein, 'Other Thought-worlds', p. 335; see also Barry Windeatt, 'Courtly Writing', in *A Concise Compendium to Chaucer*, ed. Corinne Saunders (Oxford: Blackwell, 2006), pp. 90–110, pp. 95, 100.

[32] Knight, 'Classicizing Christianity', p. 149.

[33] See the analysis of the *Miller's Tale* in Boenig, *Chaucer and the Mystics*, pp. 50–58; see also Robert E. Kaske, 'The "Canticum Canticorum" in the "Miller's Tale"', *Studies in Philology* 59, no. 3 (1962): pp. 479–500.

[34] As to a discussion on the causes and the origins of dreams, see Spearing, *Medieval Dream Poetry*; Kruger, *Dreaming in the Middle Ages*; Koonce acknowledges that the catalogue progresses from inward states to exterior spiritual forces, *Chaucer and Fame*, pp. 46-49. Delany defines the Proem as a 'plethora of

to phenomena belonging to an enhanced state of consciousness, listed in accordance with the mystics' hierarchical levels.

> God turne us every drem to goode!
> For hyt is wonder, be the roode,
> To my wyt, what causeth *swevenes*
> Eyther on morwes or on evenes,
> And why th' effect folweth of somme,
> And of somme hit shal never come;
> Why that is an *avision*
> And why this a *revelacion*,
> Why this a *drem*, why that a *sweven*,
> And noght to every man lyche even;
> Why this a *fantome*, why these *oracles*
> I not; but whoso of these *miracles*
> The causes knoweth bet then I,
> Devyne he, ... (*HF*, ll. 1–14, emphasis added)[35]

Even though the dreamer declares himself not to be knowledgeable about dream theories, the opening lines follow Macrobius's classification that the Middle Ages inherited through his *Commentary on the Dream of Scipio*. Macrobius classified five types of dreams, two being false, *phantasma* and *insomnium*, whereas the others, *visio*, *oraculum* and *somnium*, were deemed meaningful. *Visio* shows events as they will occur; in *oraculum* an authority of some kind communicates truth to the dreamer; *somnium* needs interpretation as its meaning is expressed in a figurative, symbolic manner.[36] 'Dreme', 'swevene', 'avision', 'fantome', and 'oracle' derive from Macrobius's classification; 'revelation', from the Christian tradition, was listed in Gregory the Great's (540–604) classification; 'miracle' referring to all of these 'wondrous' phenomena, also retains the significance of a supernatural event performed by God or by saints.[37]

---

contradictory information about dreams', *Chaucer's House of Fame*, p. 41; J. S. Russell defines the Proem 'errant, quirky' and introduces a 'confused persona', *English Dream Vision*, p. 178.

[35] All quotations are from Larry D. Benson, ed., *The Riverside Chaucer* (3rd edition Oxford: Oxford University Press, 1988)

[36] Ambrosius Theodosius Macrobius, *Commentarii in Somnium Scipionis*, ed. and trans. N. Moreno (Milano: Bompiani, 2007), Book I, 3.2, 3.3.

[37] Gregory the Great, *Morals on the Book of Job* (Oxford: John Henry Parker; London: F. and J. Rivington, 1844), 24.42, p. 448. 'Miracles' are defined as 'a

The following lines refer to the practices conducive to the contemplative state, commencing with those concerning the taming of the body.

For to gret febleness of her brayn,
By abstinence or by seknesse,
Prison-stewe or gret destresse (ll. 24–26)

Abstinence can refer to food deprivation as well as purity of the body attained by avoidance of sexual intercourse. Sickness is a pivotal point in mystics' lives, a watershed separating their secular activity from the subsequent state of withdrawal.[38] Illness can be overcome through a visionary episode which leads the secular person towards withdrawal from the world into a contemplative life characterised by isolation. For monastics and anchorites it entails living in a confined space. One might suggest that this is the significance of 'prison-stewe' (l. 26).[39] The discourse, then, moves on to the three steps towards contemplation.

That som man is to curious
In *studye*, or melancolyous,
Or thus so inly ful of drede
That no man may hym bote bede;
Or elles that *devocion*
Of somme, and *contemplacion*
Causeth suche dremes ofte (ll. 29–35, emphasis added)

The terms 'studye', 'devocion' and 'contemplation' closely mirror the triad theorised by the Victorines, although here 'devocion' substitutes for meditation. 'Studye' also conveys the meaning of a mental state of a particular intensity: 'a state of deep thought, meditation, contemplation' and it renders the Old French 'cogitaciun' in at least one instance: 'Sum

---

wondrous phenomenon or event', see *Middle English Dictionary* (*MED*), 1 (a), e 1 (b). http://quod.lib.umich.edu/m/med/ [accessed November 2014].

[38] It is the case of Julian of Norwich, *Shewings*, Chapter 2; for sickness as a common pattern in mystics' lives, see Wolfgang Riehle, *The Middle English Mystics*, trans. B. Standring (London: Routledge, 1981), pp. 28–29; see also Mircea Eliade, *Myths, Dreams, and Mysteries, the Encounter between Contemporary Faith and Archaic Reality* (London: Collins, 1968), pp. 74–75.

[39] See *MED*, 'prisoun'. 'Steu(e)' (n. 2) 1. (c) has the meaning of 'a small room, closet'.

[dreams] come þurgh grete stody [F cogitaciun]'.[40] *Cogitatio* is the Victorines' first step towards *contemplatio*. The second term, 'devocion' (l. 33), appears signifying 'the profound religious emotion of awe, reverence, adoration', a feeling that permeates the entire mystical experience throughout the ascent.[41] Finally one might wonder how contemplation can cause dreams (l. 34–35). I contend that the Middle English term 'drem' embodies a wider significance than the modern 'dream', 'phenomenon occurring during sleep'. It refers to a state of more profound awareness that can also take place in waking and would encompass states of consciousness ranging from the poetical creative to the contemplative state.[42] Mystics depict contemplation as 'a maner of slep', 'a lityl slomeryng', a kind of slumber.[43] The 'drems' caused by contemplation would refer to the ecstatic state of consciousness and the images contemplated during such a state.[44]

So far the excursus has progressed from abstinence to contemplation. The following lines mentioning lovers have been interpreted as referring to earthly lovers. Instead I argue that once located in the realm of the contemplative stage, the discourse remains located in the upper spheres.

---

[40] *MED*, 'studi(e) (n.) 4 (b), a1400(c1303) Mannyng *HS* (Hrl 1701) 399. http://quod.lib.umich.edu/cgi/m/mec/med-idx?type=byte&byte=198667134&egdisplay=open&egs=198697073 [accessed 29 August 2016]. *Cogitatio* is also enumerated by Gregory the Great within the classification of dreams, *Morals on the Book of Job*, 24.42, p. 448.

[41] *MED*, 'devocioun' (n.), 1.

[42] The significance of the Middle English term 'drem' would require a cross-disciplinary and transhistorical investigation that goes beyond the scope of the present article.

[43] Margery Kempe, *Book*, ch. 85, ll. 4911; 4940–41.

[44] For a correspondence between dreams and contemplation, see Hugh of St Victor, *Adnotatiunculae elucidatoriae in Joelem prophetam*, *PL* 175, col. 357, available at http://www.documentacatholicaomnia.eu/04z/z_1096-1141__Hugo_De_S_Victore__Adnotaniunculae_Elucidatoriae_In_Joelem_Prophetam__MLT.pdf.html [accessed October 2014], wherein a tripartite system including visions, dreams, and prophecy (each of which is further subdivided into three subcategories) can be found. For contemplation, also defined as *alienatio*, intertwined with dreams, see Richard St Victor, *Benjamin major*, 5,1, *PL* 196, col. 171.

Or that the cruel lyf unsofte
Which these *ilke* lovers leden
That hopen over-muche or dreden,
That purely her impressions
Causeth hem avisions (ll. 36–40. Emphasis added)

Terminology and imagery of earthly and divine love overlap. As pointed out earlier, mystics are lovers of the divine, thus the mention of lovers here could relate to contemplatives. Furthermore, one might question why Chaucer would choose to move from contemplation to worldly love. If the expression 'ilke lovers', 'these same lovers', connects to those who see oneiric images in their contemplation, then Chaucer is pointing to the lovers of the divine. Remembering that the mystic's ascent is an act of love towards the divinity, here 'avisions' (l. 40) would maintain the religious significance.[45] As to 'drede' (ll. 31; 38), one could hypothesize a twofold significance: firstly it expresses the feeling of terror and awe experienced before the sacred; secondly it would allude to the distress of the contemplative facing the visionary event, as even saintly people could not be certain that the origin of their visions were divine because the devil's wiles were a constant threat.[46] In effect in the following lines (ll. 41–42) the mention of 'spirits' causing people's dreams at night is a reminder that any phenomenon could be triggered by either good or evil spirits. No one is immune from their influence, whether 'folk' (l. 42) or mystics. Unlike the latter, the common person is unable to differentiate between dreams caused by good or evil spirits as he/she lacks the experience of the divine. The final lines of the introductory section complete the outline of the hierarchical stages, deploying a gamut of religious expressions, including the perfection of the soul and prophecy.

Or yf the soule of propre kynde
Be so parfit, as men fynde,
That yt forwot that ys to come,
And that hyt warneth alle and some
Of everych of her aventures

---

[45] Havely renders 'visions' (l. 7) as 'precognitive dream', whereas the same term in l. 40 as 'the kind of dreams that (...) reflect[s] the preoccupations of waking life', Nick Havely and Helen Phillips, eds., *Chaucer's Dream Poetry* (London: Longman, 1997), pp. 126–27, n. 39–40.

[46] Julian of Norwich, *Shewings*, Chapters 66, 69. Margery Kempe, *Book*, Chapter 59. For a discussion on the diffusion of meditative techniques to lay people, see Newman, 'What Did It Mean to Say "I Saw"?', pp. 25–29.

> Be avisions or be figures,
> But that oure flessh ne hath no myght
> To understonde hyt aryght,
> For hyt is warned to derkly (ll. 43–51)

Having achieved perfection (l. 44), the soul foreknows and forewarns future events (l. 45) through visions and 'figures' (l. 48). Although knowledge is shrouded in the mystical cloud, the perfect soul can have an insight of things to come: a clear allusion to prophecy, which was the highest stage as illustrated by the Victorines and Aquinas amongst others.[47] In Richard of St Victor's tripartite system, prophecy was the highest experience before visions and dreams.[48] Nonetheless, these visions (l. 48) cannot be grasped by intellect, they remain ineffable, as the vision is too 'derkly' (l. 51). This last term, which concludes the review of the ascending journey of the soul, is charged with the significance bestowed on it by *Corinthians* I, 13.12.[49]

On close analysis the introductory lines, besides adumbrating the medieval dream-lore, mirror the mystical discourse of the journey of the soul to God which ranges from the initial phases rooted in physicality that must be overcome, through to contemplation and prophecy, accessible to only the chosen few. When read with the mystical experience and the relevant terms used in mind, the opening section of the *HF* reveals a knowledge on Chaucer's part of the hierarchical stages as understood by mystical writers. Further, it anticipates and echoes the dreamer's path, as narrated in the rest of the poem.

Similarities between the mystics' journey and the poet's oneiric progress are evident in Books II and III. The dreamer's gradual ascent is signposted by oneiric imagery shared both by mystics and poets. After the narration of the *Aeneid*, the dreamer leaves the temple, crossing the door: 'wiket' (l. 477; 'dores' l. 480) into a desert-like landscape.[50] At this

---

[47] On prophecy, see Thomas Aquinas, *Summa Theologiae*, II, II, QQ. 171–75. http://www.corpusthomisticum.org/sth3171.html [accessed November 2014].

[48] See note 44 above.

[49] 'For now we see through a glasse, darkely', http://www.kingjamesbibleonline.org/1611-Bible/book.php?book=1+Corinthians&chapter=13&verse= [accessed November 2014].

[50] For the meaning of the desert, see Howard R. Patch, 'Chaucer's Desert', *Modern Language Notes* 34, no. 1 (1919): pp. 321–28; John M. Steadman, 'Chaucer's "Desert of Libye", Venus, and Jove (*The Hous of Fame*, 486–87)',

desolate sight, he bursts into an invocation to Christ asking him for protection from 'fantome and illusion' (l. 493).[51] Both terms point to false dreams perhaps caused by evil spirits, from which the dreamer seeks defence by addressing this prayer to the divinity 'with devocion' (l. 494) and by casting his 'eyen to the hevene' (l. 495), where he sees the golden eagle soaring in the sky. Book I illustrates different levels of mental concentration. The initial narration of the *Aeneid* could be interpreted as the stage of 'studye', intended as both modern English 'studying', and Middle English 'studye' corresponding to *cogitatio*. This re-reading of the *Aeneid* placed at the commencement of the poem could be paralleled with the practice of reading Holy Writ as counselled by spiritual directors. The dreamer passes through the door, a symbol of transition between states, and feels the unsettling sensation of disorientation. The prayer, one of the meditative techniques, as outlined previously, leads to the appearance of the eagle, a well-recognised symbol in mystical and visionary writing, that lifts the dreamer skyward.

Rowland observes that the eagle 'is an appropriate vehicle to transport the dreamer to another world'.[52] Lifted up by the soaring Eagle, the dreamer faints.

> For so astonyed and asweved
> Was every vertu in my heved,
> What with his sours and with my drede,
> That al my felynge gan to dede,
> For-whi hit was to gret affray. (ll. 549–53)

The loss of consciousness, due to dread (l. 551), is expressed by 'astonyed and asweved' (l. 549).[53] The evidence that the dreamer has lost

---

Modern Language Notes 76, no. 3 (1961): pp. 196–201. For the desert as an unusual image in a dream vision, see Delany, *Chaucer's House of Fame*, p. 58.

[51] For 'illusion' as listed by Gregory the Great as one of the causes of dreams, see *Morals on the Book of Job*, 24.42, p. 448. For an analysis of the meaning of 'fantom', see Sheila Delany, '"Phantom" and the *House of Fame'*. *The Chaucer Review* 2, no. 2 (1967): pp. 67–74.

[52] Beryl Rowland interprets the eagle as 'the symbol of eloquent speech' in 'Bishop Bradwardine, the Artificial Memory, and the *House of Fame'*, in *Chaucer at Albany*, ed. Rossell Hope Robbins (New York: Burt Franklin, 1975), pp. 41–62, p. 49, as quoted in Buckmaster, 'Meditation and Memory', p. 284. For the Eagle as 'a contemplative symbol of Philosophy', see John M. Steadman, 'Chaucer's Eagle: a Contemplative Symbol', *Publications of the Modern Language Association of America* 75, no. 3 (1960): pp. 153–59.

consciousness is twofold: firstly the eagle commands him to 'awake' (l. 556; 560); secondly the dreamer states '[m]y mynde cam to me ageyn' (l. 564). The use of sleep as a symbol for life on this earth as opposed to waking to the true life resonates here.[54] Remarkably 'swounyng' is one of the Middle English terms for rendering the ecstatic experience, swooning being a frequent event in dream visions during which the dreamer faints or loses consciousness.[55]

Within the dreamer's flight in Book II a number of ascents can be identified, commencing with the instance when the dreamer is first grasped by the eagle. The second ascent, marked by 'upper to sore / he gan' (ll. 885–86), enables the dreamer to soar above the earth, yet still recognises the landscape below. Once more the eagle soars and the dreamer's view is now so extensive that the earth has become a mere dot, 'no more semed than a prikke' (l. 907). The single point is another mystical symbol; it represents the centrality of divinity, its immateriality, and its immeasurability.[56] At this altitude, as the eagle explains, they are flying higher than Alexander the Great, Scipio, Daedalus, and Icarus (ll. 910–20). The eagle invites the dreamer to turn his face upwards (l. 925) and behold the sphere of the heavens. If the dreamer 'cast up' (l. 935) his eyes he can see as far as the Galaxy (l. 936). Thus his perspective has now radically shifted from the earth to the infinite of the cosmos; from microcosm to macrocosm. This ultimate phase of ascent of the Book fills the dreamer with pure joy.

---

[53] Referred to the senses or mental faculties 'astonyed' signifies: 'dulled, benumbed, deadened'; whereas referred to persons it means 'stupefied, stunned; unconscious', *MED*, 'astoned', 4, a; and 1. 'Asweved' signifies 'put to sleep, dulled'; *MED*, 'asweved'.

[54] For waking from the sleep of sin, see Gregory the Great, *Moralium Libri sive Expositio in Librum Beati Job*. Pars I, Lib. VIII, *PL* 75, cols. 509–1162, col. 813 C. For the three symbolic meanings of sleep and reference to the *Song of Songs*, see Gregory the Great, *Moralium Libri*, Pars I, *PL* 75, Lib. V, col. 708D, at http://www.documentacatholicaomnia.eu/04z/z_0590-0604__SS_Gregorius_I_Magnus__Moralium_Libri_Sive_Expositio_In_Librum_Beati_Job._Pars_I__MLT.pdf.html [accessed October 2014]; Gregory the Great, *Morals on the Book of Job*, pp. 282–83.

[55] For an analysis of the terms expressing the ecstatic experience, including 'excess', 'ouerpassing', and 'ravishen' used by Middle English mystics, see Riehle, *Middle English Mystics*, pp. 92–96.

[56] Julian maintains: 'And after this I saw God in a poynte', *Shewings*, XI, l. 427; Julian of Norwich, *Revelations of Divine Love*, trans. C. Wolters (London: Penguin, 1966), note p. 80.

He gan always upper to sore,
And gladded me ay more and more,
So feythfully to me spak he. (ll. 961–63)

Besides 'to entertain, amuse', 'gladen' (l. 962) also denotes comfort achieved in spiritual circumstances – indeed Rolle depicted the joy of contemplation as 'gladeness'.[57] During the experience of flying above the demons of the air, the clouds, the mists, storms, rain, and winds, the dreamer remembers two noble antecedents: Boethius and St Paul. From the former he quotes the image of Lady Philosophy clothed with wings that enables human thought to soar in the sky.[58]

And thoo thoughte y upon Boece,
That writ, "A thought may flee so hye
Wyth fetheres of Philosophye,
To passen everych element,
And whan he hath so fer ywent,
Than may be seen behynde hys bak
Cloude" – and al that y of spak. (ll. 972–78)

As with St Paul, the dreamer shares the doubt concerning his state of consciousness during his ecstatic experience: the dreamer wonders whether he is there 'in body or in gost' (l. 981). He knows not, but God does, echoing St Paul's words.[59] Emphasis placed on the parodic aspect of this circumstance has not been lacking.[60]

Once deposited by the Eagle, the dreamer learns that the loud noise coming from the House of Fame is composed of all the words spoken on the earth, which here take on a similar shape to the person who had uttered them. This personification has been viewed as a 'spoof on the general

---

[57] *MED*, 'gladen', 2, a. 'To gladden (persons or creatures); make joyful, fill with joy or bliss', *MED*, 'gladen', 1a; and 'to cheer or comfort (the heart, soul, spirits, etc.)', *MED*, 'gladen', 1b, (a).

[58] Boethius, *De consolatione philosophiae*, trans. O. Dallera (Milano: Bompiani, 1975), IV, met. 1. 1–7.

[59] Cf. 2 *Corinthians* 12.2. Saint Paul, caught up to the third Heaven, hears words that as a mortal he is not allowed to repeat.

[60] Delany, *Chaucer's House of Fame*; whereas Koonce, *Chaucer and Fame*, maintains an allegorical interpretation. According to Boitani with the reference to St Paul, Chaucer is 'tempted by mysticism', but then he does not choose it, see *Chaucer and the Imaginary World of Fame* (Cambridge: Woodbridge, Boydell & Brewer, 1984), pp. 197–98.

resurrection of the dead at Judgment Day'.[61] However, though this comment forms part of the secular interpretation of the poem, paradoxically the idea itself of the judgment day cannot be dissociated from the otherworld. Geffrey then climbs 'up the hil' (l. 1165) to the House of Fame, which stands 'upon so hygh a roche' (l. 1116). At its summit the beauty of the castle is so wonderful that 'it astonyeth yit my thought, / [a]nd maketh al my wyt to swynke' (ll. 1174–75), the dreamer is stupefied and unable to describe it, as his 'wit ne may me suffise' (l. 1179). The niches in the castle pinnacles are crowded with story-tellers and 'many thousand tymes twelve' (l. 1216) minstrels and musicians. To enter the palace he crosses a further gate, made of gold – 'a further reminder that the castle (…) is out of this world'.[62] Crossing the threshold, here again deployed as a transitional symbol, into Fame's hall, Geffrey catches sight of Fame on her throne. Her height changes, as her feet touch the earth whilst her head reaches the sky.[63] Around her throne the music has become 'hevenyssh melodye' (l. 1395) and it is so intense that it makes 'al the paleys-walles ronge' (l. 1398), resound.[64]

The last place the dreamer sees is the spinning wicker house, 'Domus Dedaly', 'Laboryntus' (l. 1920–21) as named by Chaucer.[65] The image of the whirling house is probably the most fascinating and thought-provoking challenge the dreamer-poet offers the audience. The eagle states that the dreamer cannot enter the house without its assistance, so fast does it spin.

> That but I bringe the therinne,
> Ne shalt thou never kunne gynne
> To come into hyt, out of doute,
> So faste hit whirleth, lo, aboute (ll. 2003–2006)

Whirling houses are present in romances, and entering them requires the presence of a guide, usually an animal.[66] While much has been written

---

[61] Windeatt, 'Courtly Writing', p. 101.

[62] Bennett, *Chaucer's Book of Fame*, p. 125.

[63] See Boethius, *Consolatione*, 1. pr. 1.4–42.

[64] Rolle could hear heavenly music during contemplation, see Rolle, *The Mendynge of Lyfe*, p. 19.

[65] This oneiric construction has been defined by criticism as House of Rumour, although it is never defined as such in the poem.

[66] See Wilbur O. Sypherd, *Studies in Chaucer's Hous of Fame* (1907; repr. New York: Haskell House, 1965), pp. 144–51, 173–81, on revolving houses, and p. 86–95 on the eagle as a guiding animal. Bennett briefly mentions the supernatural

concerning the sources for this structure and for the eagle in classical works, its connection with the other world has been overlooked.[67] The spinning house is extremely significant within the analysis of transitional states towards an enhanced consciousness, as it can be deemed to represent the other world. As Coomaraswamy argues: 'the other world can be regarded either as itself a revolving castle or city, or as a castle provided with a perpetually closing or revolving door'.[68] The whole rotating structure evokes, and would have a similar function as, the clashing rocks of Greek mythology, the Symplegades.[69] There the hero has to pass through these doors in a flash, as they move amazingly fast.[70] If Chaucer's whirling house belongs to the same tradition, this would explain why the dreamer is unable to enter without the eagle's aid. Tracing back the possible sources of the poem, McTurk draws attention to similar features of the *HF* and of the story in *Skáldskaparmál* narrating the theft of the poetic mead by Oðinn, in that both stories are connected to poetry, involve a journey to the otherworld, and an eagle plays a crucial role. Moreover, he draws a parallel between the function of this bird and the eagle, in Hindu mythology, which brings the *soma*, the drink conferring the gift of poetry, from heaven to earth.[71] From the perspective of the dreamer-poet, entering the wicker house represents a transition into another world, into the elevated state of poetic ecstasy.

Conclusion

Through the present analysis I hope to have demonstrated how the *HF* proceeds along the path of the mystical journey of the soul. The sense of ascent, consistent throughout the work, would be even more conspicuous

---

aspect of the whirling house and refers to two examples: *Fled Bricrend* and *Arthur of Little Britain*, in Bennett, *Chaucer's Book of Fame*, pp. 169–70.

[67] Ovid, *Metamorphoses*, 12, 39–63, where the House of Fame is described, http://www.thelatinlibrary.com/ovid/ovid.met12.shtml [accessed November 2014].

[68] Ananda K. Coomaraswamy, 'Symplegades', in *Studies and Essays in the History of Science and Learning offered in Homage to George Sarton on the Occasion of His Sixtieth Birthday, 31 August 1944*, ed. Montagu M.F. Ashley (New York: Henry Schuman, 1947), pp. 463–88, p. 480.

[69] Sypherd mentions the association of Chaucer's revolving house and the Symplegades, *Studies in Chaucer's Hous of Fame*, p. 173.

[70] Coomaraswamy, 'Symplegades', p. 481.

[71] Rory McTurk, *Chaucer and the Norse and Celtic World* (Aldershot and Burlington, VT: Ashgate, 2005), pp. 25, 28–31.

should the editorial division into three books be put aside.[72] After the classification of dreams and a synthesis of transitional states in the Proem, the temple of Venus, the crossing of the first gate, the solitude of the desert, the flight, the House of Fame, the golden gate, and heavenly music, eventually with the guidance of the eagle the dreamer enters the whirling house, the other world. Given the complexity of the poem, and the multitude of possible interpretations, the *HF* is open to continuing debate and research. In my opinion its conclusion, or the lack of it, would be rewarded by an interpretation from the mystical perspective.[73] From the abrupt conclusion of the poem and the quizzical appearance of the Man of Authority, I wish to move the focus onto the silence of the dreamer, as a traveller into the beyond. His silence would find a rationale if paralleled with the contemplative state to which the mystics aspire. In this context the dreamer spans all the steps from physical control of the body to the ineffability of the ultimate stage: the dreamer-poet's silence. Thus, through the lens of mystical perspective of the ascent, it is debatable whether the poem is truly unfinished. Its supposed incompleteness would not be the lack of a conclusion but rather the absence of words characterizing contemplation: the impossibility of putting into words the 'figures', in Chaucer's terms, appearing within the vision. The poet's silence in essence translates this sense of ineffability. As the Man of Authority's lack of words parallels the poet's silence, so the absence of any depiction of this character, corresponds to the imageless visions to which the mystics aspire. Finally, the choice of the dream structure is neither casual nor a mere literary framework. The medieval literary 'drem' is much more than a series of incongruous images reproducing the experience occurring during physiological sleep. The literary dream vision is a figurative image of the intense moment of awareness of the poet through which the creative process is achieved. Poetry and its creation during a state of enhancement similar to that of contemplation conflate the material and the supernatural, Heaven and Earth.

---

[72] Willam Joyner has challenged the division of the poem into three books, see 'Parallel Journeys in Chaucer's *House of Fame*', *Papers on Language and Literature* 12, no .1 (1976): pp. 3–19.

[73] Boenig maintains that Chaucer had planned the work as a fragment, including the lack of an ending, in 'Chaucer's *House of Fame*, the Apocalypse, and Bede', *American Benedictine Review* 36 (1985): pp. 263–77, p. 275.

Thunderbolt:
Shaping the Image of Lucifer
in the Cinquecento Veneto

Edina Eszenyi

Abstract: Lacking a clear Scriptural base, medieval authors and illuminators often related the Fall of the Angels to natural phenomena. Ancient beliefs were brought into the source discussion by Early Modern authors, among them Vincenzo Cicogna (ca. 1519–ca. 1596) from the influential Church reformer circle of the Veronese bishop Gianmatteo Giberti (1495–1543). Cicogna argued that the thunderbolt would be a proper metaphor for the fallen Lucifer, but his ideas were unwelcomed by the Inquisition.

A marriage of Heaven and Earth brought about by evil, a sin prior to the Original Sin, the Fall of the Angels holds an obscure spot in Christian theology. Angels and demons under various forms and names have been paralleled since ancient times. Belief in the expulsion of rebelling angels from heaven rationalized the presence of evil in the world and offered, at the same time, hope that it can be overcome. Relating sin and fight to angels gave a new dimension to their unquestionably good, yet perhaps somewhat flat, character. It made them a little more human. No wonder the Fall of the Angels made an impressive career in cultural history, and a treasure in the collections of the Los Angeles Getty Research Institute is evidence that it did not remain untouched by the 1500s rediscovery of the Greco-Roman cultural inheritance.

The Fall Of The Angels
When was the link between angels and demons created? The very concept of Christian angels owed much to Plato's *daimon*s, arguably the most influential pre-Christian idea related to the field. Martyr literature later slowly made spiritual companions hand over their caretaker positions to

Edina Eszenyi, 'Thunderbolt: Shaping the Image of Lucifer in the Cinquecento Veneto', *The Marriage of Heaven and Earth*, a special issue of *Culture and Cosmos*, Vol. 20, nos. 1 and 2, 2016, pp. 167–79.
www.CultureAndCosmos.org

the holy man.[1] Flanked by the cult of Saint Michael the Archangel as conqueror of evil, however, as well as textual sources such as the apocryphal Testament of Solomon, the notion that angels provide help against demonic forces made angelology grow hand in hand with demonology.[2] Their vast unexplored territories allured the early Fathers as much as later theologians and heretics, and the 1215 Fourth Lateran Council eventually imprinted their connection to the Christian tradition.[3] The Fall of the Angels became a standard scene in manuscript illumination, and the intricate theological issue made occasional appearances in the monumental arts as well.[4]

What exactly is understood by the Fall of the Rebel Angels today? According to the Catechism of the Catholic Church, a certain number of angels sinned, under the leadership of Satan, by rejecting God on their own free choice. They were created good but turned evil. This sin of the fallen angels is unforgivable, and their evil powers are conditioned by divine providence.[5]

The Catechism calls the identification of Satan with a fallen angel a teaching of 'Scripture and the Church's Tradition', but a closer look reveals

---

[1] *Paulys Realenzyklopädie der Classischen Altertumswissenschaft begonnen von George Wissowa*, 83 vols, ed. Georg Wissowa (1918), Supplement Volume 3, cols 267–322, is still the starting point for research; Francis E. Peters, *Greek Philosophical Terms: A Historical Lexicon* (New York: New York University Press, 1967), pp. 33–34; Peter Brown, *The Cult of the Saints: Its Rise and Function in Latin Christianity* (Chicago: University of Chicago Press, 1982), pp. 50–61.

[2] For an overview of the Archangel's cult: André Vauchez et al., *Culte et pèlerinages à Saint Michel en Occident: les trois monts dédiés à l'archange* (Rome: École Française de Rome, 2003); D. C. Duling, 'Testament of Solomon. New Translation and Introduction', in *The Old Testament Pseudepigrapha*, Vol. 1, ed. James H. Charlesworth, 2 vols., (London: Darton, Longman & Todd, 1983–1985), pp. 935–87.

[3] For an extensive compilation of patristic and Early Medieval angelology see 'Index *De Angelis*' in *Patrologiae cursus completus, series latina*, ed. J.-P. Migne (Turnhout: Brepols, 1956), pp. 219, 37–39; Paul M. Quay, 'Angels and Demons: The Teaching of IV Lateran', *Theological Studies* 42 (1981): pp. 20–45.

[4] From the extensive literature: Adrian Wilson and Joyce Lancaster Wilson, *A Medieval Mirror: Speculum humanae salvationis 1324–1500* (Berkeley: University of California Press, 1985).

[5] 'The Fall of the Angels', in *Catechism of the Catholic Church*, at http://www.vatican.va/archive/ccc_css/archive/catechism/p1s2c1p7.htm [accessed 28 November 2014].

that tradition plays a significantly bigger role. The Catechism refers to John 8:44 and 2 Peter 2:4, but the former passage does not mention angels at all, whereas the latter does not list Satan among the sinful angels. The Catechism argues for Satan's identification as a fallen angel also by a reference to the Fourth Lateran Council and the *Enchiridion*, the compendium of the basic texts of Catholic dogma. Both indeed declare that demons or devils were originally created good and turned evil as a result of sinning, but the key point of equalling demons with good angels before this fall is completely missing from the former and requires quite some effort to read from the latter.[6]

The lack of explicitness is less surprising in light of the fact that the Fall of the Angels has no incontestable Scriptural base. Among Biblical passages most commonly associated with the idea, Revelations 12:7-9 is maybe the best known due to its influence on the visual arts. The passage, telling the battle of angels under the leadership of Archangel Michael with a dragon-serpent called Satan, was the version that resulted in the popular imagery of Satan as serpent or dragon, often in a fight with the archangel.[7] Origen is perhaps the best known and maybe earliest among representatives of an alternative line of tradition, recognizing a fallen angel in the fallen star of Isaiah 14:12-17.[8] The Vulgate's translation of this 'morning star' as 'Lucifer' gave a name to the foremost fallen angel, a name which then blended in with Satan on the basis of further Scriptural passages such as Luke 10:18. At a certain yet undefined point, Ezekiel 28:14-17 joined the approved *loci classici* on the Fall of the Angels, as a passage which recounts the fall of a magnificent Cherub. In addition to further Biblical passages, including Isaiah 34:4 and Genesis 6:1-4, the appeal of the Fall of the Angels also greatly increased around 1260, when

---

[6] 'The Canons of the Fourth Lateran Council, 1215', in *Fordham University Center for Medieval Studies Internet Medieval Sourcebook*, at http://www.fordham.edu/halsall/basis/lateran4.asp [accessed 28 November 2014]; Heinrich Joseph Dominicus Denzinger et al., *Enchiridion symbolorum, definitionum et declarationum de rebus fidei et morum*, Nr. 800, at http://denzinger.patristica.net/enchiridion-symbolorum.html [accessed 28 November 2014].

[7] For the dragon in hagiography see Jacques Le Goff, 'Ecclesiastical Culture and Folklore in the Middle Ages: Saint Marcellus of Paris and the Dragon', in *Time, Work and Culture in the Middle Ages* (Chicago: The University of Chicago Press, 1982), pp. 159–88.

[8] 'Origen: De Principiis', *New Advent. The Catholic Encyclopedia*, at http://www.newadvent.org/fathers/0412.htm [accessed 28 November 2014].

the *Legenda Aurea* identified Archangel Michael as the leader of good angels who cast rebel angels out of Heaven.[9]

The origins and development of the tradition are difficult to trace with any considerable exactness, but the varieties clearly show that there is no shortage in Scriptural passages understandable as references to the Fall of the Angels. No single passage expresses indisputably, however, a *raison d'être* for a dragon-serpent Satan in Heaven, that Satan is identical with Lucifer, or that Lucifer is indeed the name of a spiritual creature and not that of a star, that it is the name of a demon turned angel who sinfully fell from a created perfection following a historic battle with good angels, let alone providing a clear definition of the sin that might have caused this primordial fall.

The number of opinions concerning this complex theological issue showed a steady increase from the early Middle Ages on. Hardly any theologian failed to comment on the Fall of the Angels, which perseveres and maintains its complexity to the present day. Perhaps as a direct consequence of the lack of scholarly concord, the academic research of the tradition is less extensive and manifests primarily in comments on various aspects rather than in overarching summarizing projects. Jeffrey Burton Russell's widely cited works, for instance, followed the career of the Fall of the Angels in Apocalyptic literature, while Bernard Jacob Bamberger traced the same in the rich Jewish tradition.[10] The bulk of lesser known but not less valuable publications are well represented by the conference volume of a 2001 Tübingen symposium entitled 'The Fall of the Angels', elaborating on various aspects from arguable biblical references to the medieval understanding and symbolic connotations.[11] Despite the considerable overlap of the two fields, expertise in both angelology and demonology still remains an ambitious claim from any scholar. The difference between a more theoretical versus a practical nature, between

---

[9] Giovanni Paolo Maggioni, ed., *Iacopo da Varazze: Legenda aurea* (Florence: Edizioni del Galluzzo, 1998), p. 986.

[10] Jeffrey Burton Russell, *Satan: The Early Christian Tradition* (Ithaca: Cornell University Press, 1981); Jeffrey Burton Russell, *The Prince of Darkness: Radical Evil and the Power of Good in History* (Ithaca: Cornell University Press, 1988); Jeffrey Burton Russell, *The Devil: Perceptions of Evil from Antiquity to Primitive Christianity* (Ithaca: Cornell University Press, 1987); Bernard Jacob Bamberger, *Fallen Angels: Soldiers of Satan's Realm* (Philadelphia: The Jewish Publication Society of America, 2006).

[11] Christoph Auffarth and Loren T. Stuckenbruck, eds., *The Fall of the Angels*, (Leiden: Brill: 2004).

defining creatures as opposed to describing their activities, inevitably characterizes modern scholarship as it did our predecessors.[12] The various branches of angelology, demonology, theology, and history – just to mention a few – all make relevant additions to research of the tradition, but the Fall of the Angels remains a disputed dark territory somewhere in-between.

A Cinquecento Contribution

The blurred contours of definitions inspired creative ideas regarding the storyline and application possibilities of the Fall of the Angels. The variety is well represented by a work with a cosmological reference in the collections of the Getty Research Institute, elaborately entitled as *On the names of angels and demons found in the Divine Scriptures and explained by the Fathers, dedicated to the illustrious reverend Giulio Antonio Santori, the highest cardinal of Santa Severina, and on the Ecclesiastical Hierarchy* (*Angelorvm et daemonvm nomina et attribvta passim in divinis scriptvris contenta ad patrvm sententiam explicata ad Illvstris et Reverendiss Ivlivm Antonium Sanctorivm Cardinalem Sanctae Severinae amplissimvm et de Ecclesiastica Hierarchia,* GRI MS 86-A866). The manuscript is a 170-folio lexicon of angels and demons in Latin, accompanied by a treatise on the parallel of the angelic and ecclesiastical hierarchies. It was dedicated to the distinguished Cardinal Giulio Antonio Santori (1532–1602) in the hope of eventual publication. The manuscript interprets 223 direct and metaphorical references to angels and demons, primarily from the Scriptures. The format is most reminiscent of a lexicon, with either a name or a metaphor highlighted as keywords at the top of the pages, followed by, on average, interpretations of one to two pages. The work opens with a foreword and a dedication, and then divides into two respective sections, *De Angelis* and *De Demoniis*.[13] Twelve metaphors are listed in both sections as the ones that can denote either angels or demons, and only the textual environment helps with their interpretation. Most of these metaphors are related to the Fall of the Angels. The author defined demons as fallen angels, making the disputed tradition the conceptual spine of the work.

---

[12] Armando Maggi, *Satan's Rhetoric. A Study of Renaissance Demonology* (Chicago: University of Chicago Press, 2001), pp. 21–22.

[13] I follow the word choice of the author, who used 'De Demoniis' instead of 'De demonibus'.

This unusual manuscript is signed by an Early Modern Catholic reformer author, Vincenzo Cicogna, born around 1519 into a popular local painter dynasty of Verona. Their earliest recorded member immigrated to Italy from Greece more than a century before Vicenzo's birth.[14] Vincenzo became the first ecclesiast in the family, supposedly due to his father's friendship with the local bishop Gianmatteo Giberti (1495–1543). Bishop Giberti exerted considerable influence on the theological currents of his times by way of an extensive reform process, which targeted the clergy's intellectual and moral level. His reforms served as a model even for the Council of Trent. Throughout the construction and implementation of the reforms, Giberti cooperated with a group of learned ecclesiasts with Vincenzo Cicogna among them.[15]

Cicogna established himself as a prominent ecclesiast in Verona. His networks reached as high as Charles Borromeo, whom he assisted in pastoral visits in 1564. Cicogna started to publish as rector of the local monastery of San Zeno in Oratorio, a post he held between 1544 and 1566. Prior to the *Angelorum* he published two sermon collections, one or two speeches, and a commentary of the Psalms with Cabbalistic meditations on letters of the Hebrew alphabet. The dedication to high-level ecclesiasts was a common feature of the works, and Cicogna characteristically chose an outstandingly influential ecclesiast for the *Angelorum* as well.[16] Giulio

---

[14] On the family, see primarily Raffaello Brenzoni, *Dizionario di artisti veneti* (Florence: L. S. Olschki, 1972), pp. 106–8; Gaetano Da Re, 'I Cicogna dal secolo XVI', *Madonna Verona. Bollettino del Museo Civico di Verona* 7 (1913): pp. 109–23; Luigi Simeoni, *Maestro Cicogna (1300–1326)* (Verona: Antonio Gurisatti, 1907), pp. 11–17.

[15] Angelo Turchini, 'Giberti, Gian Matteo', *Dizionario biografico degli Italiani*, Vol. 54 (Rome: Istituto della Enciclopedia Italiana Fondata da Giovanni Treccane S. p. A., 2000), http://www.treccani.it/enciclopedia/gian-matteo-giberti_%28Dizionario-Biografico%29/ [accessed 28 November 2014], with further bibliography.

[16] His earliest work, *Sermones 7* (Venice, 1556), was a collection of seven sermons on the Eucharist, dedicated to Aloysio Lipomano, bishop of Verona. The sermons were republished with six new Passion sermons in *Sermones* (Venice: Andrea Arrivabene, 1562), dedicated this time to Cardinal Marcantonio Da Mula (Amulio). The *Oratio in Bernardi Naugerii cardin*[alis] *amplissimi et episcopi veronen*[sis] *aduentu* (Venice: Iordani Zileti, 1564) was an oratory speech given by Cicogna when Cardinal Bernardo Navagero paid a visit to Verona. Jacopo Vallarsi and Pierantonio Berno, *Verona Illustrata parte seconda*, 1731, p. 422 mentions another published speech, composed at the 1565 death of the same cardinal. The same source also claims that Cicogna's latest publication, the

Antonio Santori, Cardinal of Santa Severina, was a productive man of letters, a consultant of several popes, and himself a candidate for papacy in 1592. Yet he is best known as Prefect of the Sacred Congregation of the Roman and Universal Inquisition, a post which enabled him to overview the Index of Prohibited Books.[17] This responsibility might well have been a key factor in his relationship with Vincenzo Cicogna. Cicogna's sermon commentary, the 1567 *Enarrationes in psalmos*, was '*nisi corrigantur*', included in the 1580, 1583, and 1596 Indexes; and its author imprisoned in Rome in 1573 for reasons currently unknown.[18]

The Archives of the Congregation of the Doctrine of the Faith, the congregation formerly administering the Inquisition, hold the dedication of a lost work by Cicogna, entitled *Thesaurus divina oracula et attributa continens*.[19] The 12-folio long dedication addressed Pope Gregory XIII, consequently the composition of the work is datable to 1572–1585. In all probability, the *Thesaurus* is identical with a work on prophecies and attributes mentioned in the *Angelorum*'s dedication as a manuscript that Cicogna had sent to Cardinal Santori for publication a year before the dedication was composed. The *Angelorum*'s dedication, Cicogna's literary oeuvre and biographical data point towards the year 1587 as a likely date for the *Angelorum*'s completion, and probably make it Cicogna's last work after the *Thesaurus*. This is all the more interesting as a letter attached to the *Thesaurus*' dedication prohibited Vincenzo Cicogna from publishing or even composing anything related to theology in the future.

Cicogna clearly did the opposite and more in the *Angelorum*. More than simply signalling problems with the Church with the help of angelology, he took a constructive approach and argued for the universal nature of Christianity by highlighting its understated harmony with pre-Christian

---

Enarrationes in psalmos with a dedication to Pius V (*Patauii: ex officina Laurentij Pasquati and Patauii:* 1567, 1568) appeared in print already in 1556, but I found no confirmation of this early work's claims in other sources.

[17] Saverio Ricci, *Il Sommo Inquisitore: Giulio Antonio Santori Tra Autobiografia e Storia (1532–1602)*, Piccoli Saggi 15 (Roma: Salerno, 2002), with further bibliography.

[18] Da Re, 'I Cicogna dal secolo XVI', p. 119; Enrico Maria Guzzo, 'Il palazzo Del Bene di San Zeno in Oratorio in Verona (e le relazioni di Giovanni Battista Del Bene con alcuni artisti veronesi)', in *La famiglia Del Bene di Verona e Rovereto e la villa Del Bene di Volargne*, ed. Gian Maria Varanini (Rovereto: Accademia Roveretana degli Agiati, 1996), pp. 81–113, n. 40.

[19] *Thesaurus d<ivina> oracula et attributa continens*, Archivio della Congregazione per la Dottrina delle Fede, Index *Protocolli* G, fols 306^r–317^r.

philosophical systems. His methodology is well illustrated by the presentation of the thunderbolt as a symbol of Lucifer's fall. The simile is explained in the wider context of the Fall of the Angels in the lexicon entry with the Latin keyword *Fulgur* (Fol. 123[r-v]).

Cicogna painted a multi-dimensional portrait of Lucifer as the leader of the fallen angels. References to the demon-turned-angel recur in numerous of his lexicon entries, revealing additional details to the basics set down in Lucifer's own, doubled entry. The *De Angelis* section's *Lucifer* entry (Fol. 55[r]) does not paint an overall negative image; it is a short elaboration on Lucifer's pre-eminence prior to the fall. He was as an angel similar to the morning star, brighter than any other angel, with a unique splendour similar to that of the stars. The last sentence in the entry closes with a note that a sin committed by this beautiful angel inevitably brought about his fall, and directs the reader to the *De Demoniis* section's corresponding entry.[20] The author's voice is cold and documentary, not giving away regret or empathy over the loss of the splendid creature just described. Continuing the star simile in the *De Demoniis* section's *Lucifer* entry (Fol. 134[r-v]) with a longer elaboration on the once brightest creature turning into the Prince of Darkness, Cicogna refers to Augustine and St. Jerome when identifying pride (*superbia*) and the desire for divine nature as the fallen angel's sin.[21]

While Cicogna names Revelations, Isaiah and Luke as the biblical sources of the Fall of Lucifer at this point, he focuses exclusively on the latter in the *Fulgur* entry (Fol. 123[r-v]). Luke 10:18 is phrased as 'I saw Satan fall like lightning from heaven…' in today's English Standard Bible translation, whereas the Vulgate used the Latin *fulgur* meaning 'thunderbolt': '… et ait illis videbam Satanan sicut fulgur de caelo cadentem'.[22] Cicogna lists arguments in the entry for this natural phenomenon being a particularly apt metaphor for Lucifer. He relies heavily on Pliny the Elder's *Natural History*, where the thunderbolt and the

---

[20] 'Angelus antequam e celo laberetur est appellatus ob eius singularem splendorem: qui stellae matutinae comparetur splendidiori caeteris astris: […] sed eum per superbiam a Deo defecisset, nomen quidem Luciferi retinuit, sed non rem: non enim lucem, sed tenebras et noctem affert. […] Sed de hoc de Daemonii verbo LUCIFER latius'. (GRI MS 86-A866 Fol. 55[r])

[21] 'Dei naturam et substantiam affectarent, e caelo deturbati, obtenebrati sunt, et principes tenebrarum facti […] propter superbiam suam de celo in terram praecipitatus est…' (GRI MS 86-A866 Fol. 134[v]).

[22] *The Bible Gateway Searchable Online Bible*, at http://www.biblegateway.com/ [accessed 28 November 2014].

lightning constitute a recurring theme, with Book 2 Chapter 43 covering most of the subject. Pliny used the Vulgate's *fulgur* for thunderbolt and *fulgetrum* for lightning, saying both are accompanied by a short flash of light but only the thunderbolt by a roaring noise originating from the collision of clouds.[23]

Both features had metaphorical meanings for Cicogna. In the momentarily flash he recognized a reminder of how quickly Lucifer fell when the once brightest angel turned dark. As the thunderbolt is caused by a great uproar at the collision of clouds, Lucifer's fall followed a battle with the good angels. Lightning is light only, which could only stand for Lucifer's fall without the battle of the angels. As there is no thunderbolt without lightning, the battle would not have happened without Lucifer's fall, but it must have taken place as it is recalled in Revelations 12:7-9, he argues.[24]

The battle of the angels is again a recurring theme in the *Angelorum*, with the most detailed analysis embedded in the entry on Archangel Michael (fols 56^v-57^r). Cicogna explains the archangel's name in the traditional way, with the Hebrew 'Mi Kha El' and the Latin 'Quis ut Deus' meaning 'Who is like God?', and identifies these as the words the archangel replied to a forsaken angel's desire to become equal to God. Cicogna uses the archangel's name to remind his readers of the importance of humility, by pointing out that its lack caused Lucifer to fall. He also adds that the other angels joined Michael in crying out the same words, which demonstrates the ultimate humility of angels, this supreme humility of the creatures who excel all other creatures.[25]

---

[23] Pliny the Elder, *The Natural History*, trans. John Bostock (London, 1855), http://www.perseus.tufts.edu/hopper/text?doc=Plin.+Nat.+&fromdoc=Perseus%3 Atext%3A1999.02.0137 [accessed 28 November 2014]; Trevor Morgan Murphy, *Pliny the Elder's Natural History: The Empire in the Encyclopedia* (Oxford: Oxford University Press, 2004).

[24] 'Fulgur enim eodem fere momento extinguitur, quo fit nam et Satanas vix a Deo optimus creatus, pessimus est affectus, et lux illa, in qua conditus fuerat, in tenebras est conversa: [...] Ut autem fulgur ex collisione nubium cum fragore fit: ita Santanae casus non sine bonorum Angelorum contradictione factus est. [...] Sed Christus non fulgetrum, sed fulgur Satanam appellavit, quamvis fulgur sine fulgetro non fiat'. (GRI MS 86-A866 Fol. 123^{r-v})

[25] '...cui eum adversaretur Angelus sanctus, et Sathanam superbiae argueret, hac voce, Quis sicut Deus, ab illa nomen consecutus est, et appellatus est Michael [...] Quae quidem voc cum non unius tantum Angeli sed omnium Angelorum sit, qui

The entry comparing them to clouds (*Nubes*, Fol. 62^r) further explains why Cicogna found the collision of clouds proper metaphors for the battle of angels. Angels are similar to clouds because they are light, subtle and fast, and because they mediate between heaven and earth, he says. As clouds preserve the earth from the heat of the sun, angels preserve people from evil, and the sky is their dwelling place. Thematically this entry belongs to the group explaining the angelic nature while offering one of the most picturesque metaphors in the work.[26]

After clarifying the basics, Cicogna goes on to draw a parallel between the destructive nature of the thunderbolt and evil. He lists three types of thunderstorms on the basis of *Natural History* 2.52, all very destructive. The first one is the dry thunderstorm that produces no precipitation, and dissipates objects. The second one is what he calls moist thunderbolt, meaning apparently the thunderstorm with precipitation, able to blacken objects. A mysterious third type, which Cicogna calls bright lightning (*quod clarum vocant*) after Pliny, empties vessels without trace while leaving their lids intact. This reminds Cicogna of the way the devil harms the souls, often emptying them in subtle and unnoticeable ways. The entry concludes with a reminder that as the light remains an essential element of the thunderbolt, Lucifer has the capacity and unfortunate tendency to transform himself into an angel of light and to mislead people. Finally, and maybe on a somewhat lighter note, Cicogna also recalls that the thunderbolt was superstitiously applauded in ancient civilizations: *Natural History* notes that worshipping lightning by clucking with the tongue was a worldwide tradition. Cicogna sees a craft of Lucifer himself in this superstition, an expression of worship that the fallen angel sinfully desired for himself.[27]

---

pugnabant cum Diabolo, [...] ita maximam Angelorum humilitatem demonstrat...' (GRI MS 86-A866 Fol. 56^v).

[26] 'Nubes itaque Angeli sunt appellati propter singularem eorum levitatem, subtilitatem et agilitatem, longe nubibus praestantiorem: Et ut nubis alta petunt et mediae inter celum et terram existunt: ita Angeli medii sunt inter homines et Deum. Et ut nubis solis ardorem temperant per diem, ne terra exuratur, et illam rore et pluvi salutari fecundant: ita Angeli a Diaboli saevitia et ira nos tuentur, et donis celestibus exornatos faciunt'. (GRI MS 86-A866 Fol. 62^r).

[27] 'Diaboli autem asta factum fuisse crediderim, ut omnium gentium /sicut docet Plinius/ consensu popysmis fulgetra adorentur. [...] Quae sint [missing word] /scribit Plinius/ non adurunt, sed dissipant: quae humida, non urunt, sed infuscant: Tertium est, quod clarum vocant mirificae maximae naturae, quod dolia exhauriuntur intactis operimentis nulloque alio vestigio relicto. Diabolus enim,

The use of Pliny to explain Christian theology is evidence of the harmony sought by Cicogna between Christianity and pre-Christian cosmological ideas. References to non-Christian sources are abundant in the *Angelorum,* markedly represented by Greek philosophers such as Socrates, Plato, Aristotle and Empedocles. Cicogna claims for instance that Plato was talking about angels when he compared rulers to shepherds in *Republic* 342e, because only angels can meet the requirements of the good ruler as defined by Plato.[28] Ancient philosophy serves Christianity in Cicogna's system, but does not challenge its priority. This is demonstrated, among other points, by Cicogna's comment on Aristotle calling death 'the most terrible of all things' in *Nichomachean Ethics* III.6. Cicogna argues that the devil is in fact worse than death, being its very author.[29]

Cicogna this way defines the roots of his angelology in both Christian and non-Christian tradition, which is in tremendous contrast with the actual title of the *Angelorum,* promising a simple summary of the Church Fathers' views on angels. Furthermore, and again in sharp contrast with the title, Cicogna does not introduce angels and demons from canonical writings only. The angel Raphael, for example, is introduced in detail in a separate entry in his lexicon, although Raphael appears in the deuterocanonical Book of Tobith (Fol. 71^v). The apocryphal angels Ariel (Fol. 16^{r-v}), Ieremiel (fols 49^v-50^r), Raziel (Fol. 72^{r-v}) and Uriel (Fol. 88^{r-v}) similarly receive their own entries. As opposed to pre-Christian writings, apocryphal literature is not quoted by Cicogna even though he apparently acknowledges it as a relevant source of Christian angelology. Nevertheless the reconciliation of Christian and Jewish ideas is traceable in the *Angelorum.* The angelic hierarchy, for example, is explained with the help of the Kabbalah's Sephirotic Tree on the only illustrated page of the manuscript (Fol. 6^v), where the Sephirahs are represented by schematic figures of God and the nine angelic orders. The Sephirotic Tree represents

---

quem attigerit, adurit odium dissipat luxuria, offuscat ignorantia, et nullo vestigio relicto etiam animam et virtutes omnes perdit. [...] semper et in omnibus male olet, tunc etiam cum in Angelum lucis se transformat'. (GRI MS 86-A866 Fol. 123^{r-v}).

[28] 'Nam idem Plato aureum illud speculum sub Saturno recensens Angelos pastores hominum tunc fuisse asserit: Ut enim oves non recte nisi a pastore ducuntur ita homines non recte duci possunt nisi a Deo: qui nobis tanquam duces et pastores Angelos suos esse voluit'. (GRI MS 86-A866 Fol. 65^{r-v}).

[29] 'Aristoteles extremum omnium malorum appellavit mortem: quod omnia mala superet. Sed nobis ipsa morte deterior est Diabolus, qui mortis author est et origo fuit...' (GRI MS 86-A866 Fol. 120^{r-v}).

for Cicogna the angelic contemplation and assistance of God, with higher angelic choirs forwarding divine orders to lower choirs (fols 6ᵛ-7ʳ). This way the *Angelorum* voices the very same desire for presenting Christianity as a universal umbrella above other religions that had previously been expressed in Cicogna's best known work, the prohibited *Enarrationes in Psalmos*. The understanding of a unified Church with reconciled Jews characterised the Catholic reformer Giberti Circle's approach to Catholic reformation and plausibly remained Vincenzo Cicogna's 'principal spiritual desire'.[30]

Conclusion

The Fall of the Angels brought angels down to earth both in a literal and metaphorical sense, which might explain the tradition's perseverance throughout time despite the lack of clear Scriptural sources. This feature was nevertheless disturbing for ecclesiastical authors like Vincenzo Cicogna, who tried to integrate the Fall of the Angels into biblical interpretation. Cicogna's presentation of the tradition reveals that he joined the line of those authors who made no clear preference among the Scriptural passages approved as references to it. He was seeking rather a cohesive narration in his *Angelorum*, one that weaves together all possibilities instead of refusing any that might be relevant.

His angelology is liminal: medieval in its primary source material but Early Modern in its attempt at the integration of the pre-Christian, ancient Greek, Roman, and Jewish inheritance. Cicogna refers to Pliny as his 'fellow countryman' (*conterraneus*, Fol. 26ᵛ), using the same word as Pliny did when referring to Catullus in the dedication of the *Natural History*. This emphasis on the common origins suggests that Cicogna's preference of this work over other similar sources could well have been influenced by his own Greek origins. Was his conciliatory approach due to antiquity's rediscovery in the Italy of the 1500s, or was it simply a man of Greek origins searching for his own roots? The two probably coincided in Cicogna's character, and the thunderbolt as a metaphor of Lucifer's fall gave him a good occasion to explore the depths.

---

[30] Giuseppe Conforti, 'Villa Del Bene: iconografia e inquietudini religiose nel Cinquecento. Gli affreschi della loggia e dell'Apocalisse', in *Annuario Storico della Valpolicella 2003-2004*, ed. Andrea Brugnoli (Verona: Centro di Documentazione per la Storia della Valpolicella, 2004), pp. 99–119; A. Olivieri, 'Simeone Simeoni 'filatorio' di Vicenza (1570): il dibattito su charitas e pauperes', *Quaderni di Storia Religiosa* 2 (1995): pp. 234–36.

In any case, the welcome his approach received was apparently colder than expected. The Archives of the Congregation of the Faith register a document containing Inquisitorial notes about Cicogna's *Angelorum*, by all probability corrections, made by Bishop Federicus Metius. The bishop is referred to as a *familiaris* of Cardinal Santori in numerous lists of counsellors (*consultores*) of the Index of Prohibited Books, and he is normally listed among primary counsellors.[31] His relationship with Cardinal Santori strongly suggests that the *Angelorum* managed to get attention from its dedicatee, but was not fully, or at all, welcomed. The exact nature of the inquisitorial corrections remains a challenge for further research until the now lost censorship document hopefully resurfaces one day.

Acknowledgements
I thank the organizers of the Marriage of Heaven and Earth Sophia Centre conference for the opportunity of presenting my research, as well as the audience for the constructive questions and comments.

---

[31] Archivio della Congregatione per la Dottrina della Fede, Index *Protocolli* I, fols 359ᵛ, 361ᵛ, 362ʳ, 360ʳ, 366ʳ, 373ʳ.

The Worldly Faces of the Heavens:
Nature and Seventeenth-Century English
Astrological Images

Alexander Cummins

Abstract. The astrological image magic of early modern Europe included the magical use of representations of planets, zodiacal signs, fixed stars, decans, lunar mansions and even geomantic figures. These images were utilised magically to bring about particular conditions. Along with investigating representations of occult identity and magical purpose in these pictorial forms of the stars of the heavens – considering the symbolic iconography of early modern sigils – this paper will explore the material and environmental dimensions of their construction and use: analysing alchemical notions of planetary metals as living organic matter, and interrelations between written, ideographic and iconographic astrological charms. The paper will also consider the locative aspects of consecrating, deploying, and disposing of such astrological images in contemporary magical practices: from charging love charms in brothels, to burying magical objects at crossroads and to pitching curse-dolls into rivers. Finally, this paper will explicate relations between natural components of the landscape and their representation in astrological image magic, such as the various significances of the laurel and the apple in an image of Venus. It will demonstrate that such relations between affective space and effective ritual are central to early modern occult philosophy and magical practice.

Astrological images upon sigils – magical objects or, occasionally, the markings thereon – were not simply artistic representations, but considered operative and efficacious magical objects that could draw, contain, and radiate astral virtues to influence a variety of states and behaviours in people, animals, plants, the land and fate itself. Astrological images were framed and utilised to change the world, or to at least exert influence on parts of it to suit the astrologer-magician.

This article will delineate the environmental – by which is connoted ecological, cosmological, geographical and botanical – dimensions of astrological images, in order to apprehend and analyse magical approaches

Alexander Cummins, 'The Worldly Faces of the Heavens: Nature and Seventeenth-Century English Astrological Images', *The Marriage of Heaven and Earth*, a special issue of *Culture and Cosmos*, Vol. 20, nos. 1 and 2, 2016, pp. 181–200.
www.CultureAndCosmos.org

to and engagements with the natural world. This notion of 'worldly' astrological image magic will look at the significances of time, place and flora and fauna in human affairs, concentrating on the endeavours of agriculture and enchantment.

We know of astrology's more general place in social, print, political and medical histories of astrology,[1] as well as seeing its utility in folk magic and amulet-craft, and its role in ritual magic[2] and in the manuscripts

---

[1] For solid foundations in social and print history of astrology, the exemplars are still Keith Thomas, *Religion and the Decline of Magic* (London: Oxford University Press, 1971) and Bernard Capp, *Astrology and the Popular Press: English Almanacs 1500–1800* (London: Faber, 1979). For the political dimensions of seventeenth-century English astrology, see Harry Rusche, 'Merlini Anglici: Astrology and Propaganda from 1644 to 1651', *English Historical Review* 80 (1965): pp. 322–33; Patrick Curry, *Power and Prophecy* (Cambridge: Polity Press, 1989); W.E. Burns, 'A Whig Apocalypse: Astrology, Millenarianism, and Politics in England During the Restoration Crisis, 1678–1683', in *Millenarianism and Messianism in Early Modern European Culture: The Millenarian Turn*, ed. J. E Force and R. H. Popkin (London: Springer, 2001). For a history of astrological medicine, see Michael MacDonald, *Mystical Bedlam: Madness, Anxiety and Healing in Seventeenth Century England* (Cambridge: Cambridge University Press, 1981); Michael MacDonald, 'The Career of Astrological Medicine in England', in *Religio Medici: Medicine and Religion in Seventeenth-century England*, ed. O.P. Grell and A. Cunningham (Aldershot: Scolar Press, 1996); Lauren Kassell, *Medicine and Magic in Elizabethan London: Simon Forman – Astrologer, Alchemist, and Physician* (Oxford: Oxford University Press, 2005).
[2] See Charles Burnett, 'Talismans: Magic as Science? Necromancy among the Seven Liberal Arts' in Charles Burnett, ed., *Magic and Divination in the Middle Ages: Texts and Technicians in the Islamic and Christian Worlds* (Aldershot: Variorum, 1996), p. 1–15; Richard Kieckhefer, *Forbidden Rites: A Necromancer's Manual of the Fifteenth Century* (University Park, PA: Pennsylvania State University Press, 1998); Stephen Wilson, *The Magical Universe: Everyday Ritual and Magic in Pre-Modern Europe* (London: Hambledon & London, 2004); Don C. Skemer, *Binding Words: Textual Amulets in the Middle Ages* (Pennsylvania: Penn State Press, 2006); Owen Davies, *Popular Magic: Cunning-folk in English History* (New York: Hambledon Continuum, 2007); Owen Davies, *Grimoires: A History of Magic Books* (Oxford: Oxford University Press, 2009).). For the role of astrology in early modern manuscript copies of older grimoires, see MS Sloane 3826 collected in *Sepher Raziel: Liber Salomonis, A Sixteenth-Century English Grimoire*, ed. Don Karr and Stephen Skinner (London: Llewellyn Publications, 2010) and *The Goetia of Dr Rudd: The Angels & Demons of Liber Malorum Spirituum Seu Goetia Lemegeton Clavicula Salomonis: with a Study of the*

and workbooks of various kinds of early modern magicians.[3] While contemporary debates about the exact sphere of influence or effects of astrology arguably became increasingly heated in the early modern period, it can nevertheless still be said 'belief in natural astrology was almost universal in Tudor England'.[4] During the seventeenth century, its reach and influence achieved unprecedented heights.[5] This was in part due to the lifting of restrictions on printing presses in the middle part of the century, leading to 'several years of virtually complete freedom of the press'.[6] This facilitated an 'explosion' in printed almanac materials, translations of

---

Techniques of Evocation in the Context of the Angel Magic Tradition of the Seventeenth Century, ed. Stephen Skinner and David Rankine (London: Golden Hoard Press, 2007), which collects 'Liber Malorum Spirituum seu Goetia' from MS Harley 6483 with MS Harley 6482, MS Sloane 3824 and MS Wellcome 3203.

[3] See the decrypted diary entries in C. H. Josten, ed., *Elias Ashmole (1617–1692): his autobiographical and historical notes, his correspondence, and other contemporary sources relating to his life and work* (Oxford: Oxford University Press, 1966) and Michael Hunter and Annabel Gregory, eds., *An Astrological Diary of the Seventeenth Century: Samuel Jeake of Rye, 1652–1699* (Oxford: Oxford University Press, 1988). For a transcription of a seventeenth-century magical practitioner's working notebook of charms, conjurations, and prayers, along with tables of correspondences, recipes and copied sections from grimoires and books of occult philosophy, see David Rankine, ed., *The Grimoire of Arthur Gauntlet* (London: Avalonia, 2011), which presents and annotates MS Sloane 3851.

[4] See D. C. Allen, *The Star-Crossed Renaissance: The Quarrel about Astrology and its Influence in England* (London: Psychology Press, 1967); Richard Dunn, 'The Status of Astrology in Elizabethan England 1558–1603' (PhD thesis, University of Cambridge, 1992); Brian Copenhaver, 'Natural Magic, Hermetism, and Occultism in Early Modern Science', in *Reappraisals of the Scientific Revolution*, ed. David Linberg and Robert Westman (Cambridge: Cambridge University Press, 1990), pp. 261–301. For specific debate around astrological images, see Martha R. Baldwin 'Toads and Plague: The Amulet Controversy in Seventeenth-Century Medicine', *Bulletin of the History of Medicine* 67 (1993): pp. 227–47; Anna Marie Roos, 'Israel Hiebner's Astrological Amulets and the English Sigil War', *Culture and Cosmos* 6, no. 2 (2002): pp. 17–43; Capp, *Almanacs*, p. 31.

[5] For more on seventeenth-century English astrology and its environmental, political and social functions – as well as its ties with occult philosophy and magical practice – see Alexander Cummins, *The Starry Rubric: Astrology and Magic in Seventeenth-Century England* (Milton Keynes: Hadean Press, 2012).

[6] Christopher Hill, *Some Intellectual Consequences of the English Revolution* (London: University of Wisconsin Press, 1980), p. 7.

foreign texts, and print copies of occult manuscripts that had been
circulating around Europe. Seventeenth-century English folk learned their
astrological magic from handbooks, almanacs, herbals and kitchen physic
manuals, as well as treatises on natural magic and occult philosophy, and
even a few grimoires of ritual magic.[7]

Seventeenth-century England is also an important period to examine in
the history of astrological magic, as we possess a variety of records – from
both personal and professional correspondence, as well as private diary and
journal notes and even patient case-files – of particular sets of affiliations
of magical and astrological practitioners sharing information, teaching
their trade, and passing on their papers and their magical objects.[8] The
most central figures in these cross-generational seventeenth-century
'networks' are Simon Forman (1552–1611), Richard Napier (1559–1634),
William Lilly (1602–1681) and Elias Ashmole (1617–1692), but the wider
lines of transmission and influence clearly extended out further to many
other practitioners. As but two pertinent examples of such networking, we
know the highly regarded Arabic manual of astrological magic, the
Picatrix, was shared, copied, studied and passed on between practitioners;[9]

---

[7] For handbooks, see, for example, William Lilly, *Christian Astrology* (London,
1647) and Henry Coley, *Clavis Astrologiae Elimata: or a Key to the Whole Art of
Astrology New Filed and Polished* (London, 1676). Almanacs ranged from mainly
calendrical pamphlets of 'common notes and moouable Feasts', such as Daniel
Browne, *A New Almanacke* (London, 1620), to polemical political and theological
tracts, like John Booker, *The Bloody Almanack* (London, 1642). For herbals and
physics manuals, see Nicholas Culpeper, *The English Physician* (London, 1652)
and Joseph Blagrave, *Astrological Practice of Physick* (London, 1671). For occult
philosophical treatises, see Jean Baptiste Porta, *Natural Magick* (London, 1658)
and, most importantly, Heinrich Cornelius Agrippa, *Three Books of Occult
Philosophy*, ed. Donald Tyson (London, 1651). The grimoire of this period most
relevant to astrological magic appears to be the *Heptameron*, a popular system of
planetary angel conjuration, contained in the Robert Turner translation of Heinrich
Cornelius Agrippa's *Fourth Book of Occult Philosophy* (London, 1665). .
[8] See Kassell, *Medicine and Magic*, pp. 215, 229; MacDonald, *Mystical Bedlam*, p.
213; Josten, *Ashmole*, pp. 1208, 1454–55, 1663–64.
[9] In 1592, for instance, Forman made his own copy of the *Picatrix* (MS Ashm 244,
fol. 97), and on 5 January 1648, Ashmole records (in cipher) delivering a copy of
the *Picatrix* to William Lilly. (MS Ashm 1136, fol. 184). We also have records of
Forman's notes from *Picatrix* in his writings and calculations (MS Ashm 244, fols.
45, 97; Ashm 431, fols. 146–146v; Ashm 1491, p. 1128), including quoting it in
his biblical commentary on Genesis, alongside Augustine and others (MS Ashm

and that in 1611 Forman sent Napier some of his brass moulds with which to cast astrological sigils, along with instructions for use and reflections on practice.[10]

In his essay on the 'Heavens' (from a ca. 1606–8 manuscript), the astrologer-physician and magician Simon Forman defined: 'An ymage is the force of coelestialle bodies flowinge and soe ymages worke by vertue and similitude'.[11] Similitude as an operational principle of image magic is predicated on the axiom that a representation holds some of the power, effect or influence of the represented. This notion was itself derived from observing how not only the shape but also the markings of natural materials intimated their affective and effective applications. This itself became an argument for the natural and potentially pious use of images: in

'the use of such Characters, Letters, Words, Figures, &c. Formed or Insculpted upon any Matter we make use of, we are led to it by the president [i.e. precedent] of Nature, who Stamps most notable and marvellous Figures upon... Plants, Rootes, Seeds, Fruits, nay even upon rude Stones, Flints, and other inferior Bodies.'[12]

These signatures of natural materials – these impressions and marks that can be read and interpreted as part of the Book of Nature's instructions for use – 'are the Characters and Figures of those Starrs, by whom they are principally governed, and with these particular Stamps, have also peculiar and different vertues bestowed upon them'.[13]

So, the image of an angry man 'of a great body, with reddish eyes, and great strength', cast in metals and other materials – by being an image associated with fiery Aries and therefore acting as a catalyst in drawing, containing and radiating such astral virtues – 'signifieth and causeth

---

802, fols. 3–12, 1–2; for further on the unusual foliation of this rebound document, see Kassell, *Medicine and Magic*, p. 194 no. 21), and drawing on this text in his own expositions of natural magic, alongside Hermes, Paracelsus and others (MS Sloane 3822, fols. 68–75; MS Ashm 244, fols. 35–60). For more on this reception and circulation of *Picatrix*, see David Pingree, ed., *Picatrix: The Latin Version of the Ghayat al-hakim*, Vol. 39 (London: Warburg Institute , 1986), pp. xix, liii–lv.

[10] MS Ashm 240, fol. 106.

[11] MS Sloane 3822, fol. 81.

[12] Elias Ashmole, *Theatrum Chemicum Britannicum* (London, 1651), p. 463–44. Emphasis added.

[13] Ashmole, *Theatrum Chemicum Britannicum*, p. 464.

boldness, fortitude, loftinesse and shamelessness'.[14] These decorated items
were made at times at which the astrological force was best situated to
share its power, consecrated with prayers, incenses and other ceremonies
of empowerment, and then activated by being placed, usually by hanging,
at the site at which their influence, protection and power was needed. They
represented, in a very real sense, a frozen piece of the influence of the
heavens: an en-mattered source of astral power.

The foundations of astrological image magic in similitude are especially
important in demonstrating how early modern astrology and magic
supported one another in a mutual interrelation of theory and praxis.
Images were made through the magical principles of election and
exposure, became contagious, and operated by a sympathetic and
antipathetic affectivity. Images were also bestowed with their virtues by
the stars and therefore also depended on thoroughly astrological
ontological categories, i.e. the powers and effects of the planetary spheres,
zodiacal signs et al.

Forman explained that sigils 'enclosed som parte of the vertue of
heaven and of the plannets according to the tyme that it is stamped caste or
engraven or written in'.[15] Such astrological 'election' was generally
performed to decide upon a good time to, say, travel, begin a project, or
even propose to one's beloved.[16] Election was also considered vital for
establishing 'proper times for surgery, letting blood or taking medicine'.[17]
Joseph Blagrave (1610–1679) was eager however to emphasise 'an exact
time must be obtained whereby to erect your Figure aright, whereby to
give judgment upon the disease, its cause and termination'.[18] Not

---

[14] Agrippa, *Three Books of Occult Philosophy*, p. 377. For more on the early
modern gendering of anger, see Gwynne Kennedy, *Just Anger: Representing
Women's Anger in Early Modern England* (Illinois: Southern Illinois University
Press, 2000). For a contemporary source on anger, see Marin Cureau de La
Chambre, *A Physical Discourse Touching on The Nature and Effects of the
Courageous Passions* (London, 1658).

[15] MS Ashm 392, f. 46.

[16] Ashmole frequently elected for journeys. For instance, having elected to travel
at 10:15am, he noted '[Pisces] ascended, [Leo] in the ascendant, [Jupiter] in his
own house... This was a journey of great pleasure and good reception from
friends' when he travelled on the 27th December 1651 from London to Bradfield.
MS Ashm 374, fol. 133; such as Samuel Jeake did for his proposal to Elizabeth
Hartshorne. Hunter and Gregory, *Jeake* p. 19.

[17] Capp, *Almanacs*, p. 64.

[18] Blagrave, *Physick*, sig. Bv. Emphasis added.

coincidentally, the importance of precise time keeping was also underlined in a treatise on sigils by Israel Hiebner (1619–1668). Indeed, Anna Marie Roos has argued this emphasis on exactitude is the key to understanding how the publication of Hiebner's *Mysterium Sigillorum* – a work dedicated to the production and utility of magical objects – was regarded as such a 'landmark event among reforming and scientific astrologers'.[19] Hiebner insisted: 'you must have an exact Watch made, that shows the Hours, Quarters, and Minutes', for the Sigil 'must be stamp'd in a Minute, because this Impression occasions the Power of the heavenly Influence; for the Heavenly Influences of the Stars are as quick and nimble, as an Arrow out of a Bow, or a Bullet out of Gun, so quick must this Impression be'.[20]

Certainly Ashmole seemed to have similar convictions about precision, recording casting the astrological images to rid his house of 'Flyes, Fleas, Caterpillars & Toades', at exactly 11am, 2.17pm, 3.15pm and 4.30pm on 18 July 1650.[21] A further note informs us that the first three sets were cast as '[Saturn] and [Mars] continued in the 8: house til 4.15' P. M. Then 11.35 [Sagittarius] Ascend'.[22] We know therefore the conjunction of these planets was considered to be best harnessed in the eighth house, usually called Mors and signifying death.[23] This may have simply been the house most apt to harness the forces of malefic Mars and Saturn, or it may have had a specific bearing on the purpose for which the sigils were to be employed: killing domestic pests. Thus proper timing was accounted by clock time, by aspects of planetary bodies, and by location of conjunction within the Houses of the Heavens.

Not only were astrologers accurate with their hours and minutes, they were also accurate with the kinds of timing systems they used. Astrological chronometry used three sorts of hours. 'Natural' hours were the familiar modern type of hour using a clock, counted from midnight. 'Artificial' hours were an equal twelfth of the period of daylight (or night-time), and thus varying according to the year. These artificial hours were sometimes referred to as planetary hours, when astrologers would 'distribute each of those hours to every one to the Planets according to the order of

---

[19] Roos, 'The English Sigil War', p. 17.
[20] Israel Hiebner, *Mysterium Sigillorum, Herbarum & Lapidum* (London, 1698), p. 178.
[21] MS Ashm 431, fols. 121–22.
[22] MS Ashm 431, fol. 122v.
[23] Lilly lists it as concerning 'The Estate of Men deceased, Death, its quality and nature'. Lilly, *Christian Astrology*, p. 54.

successions'.[24] Finally, there were 'Magical' hours, which 'determined the times to make magical amulets, for instance, and were measured according to the ascension of the ecliptic line of the eighth heaven'.[25] Forman insisted that the casting of such objects must therefore be conducted according to the last of these.[26] In practice, it seems the diversity of forms of hours was used in concert. Ashmole set four different elections to determine the best time to cast a sigil of Jupiter on 5 May 1678, noting 'At 7.45 P.M. Hora [Jupiter] Art[ifical] begins, and for the following 15' both the Art[ifical] and Nat[ural] hour, is [Jupiter]... This I take to be the best time of the... times'.[27]

Early modern astrology comprehended each hour – and indeed, each kind of hour – as a window or conduit of the astral virtues presently in motion, providing an astrological undercurrent of meaning and potential affect to time itself. In examining how astrological magic situated itself within its environment, we may perhaps begin analysing these situations in the broadest sense of space and time. The condition of the heavens, both in terms of theoretical cosmological philosophy and the practical quotidian interpretation of celestial events, was arguably one of the more foundational dimensions in considering how the astrological magical image maker apprehended and operationalised their environment. Likewise, as Proclus remarked – and, significantly, as Agrippa quoted – 'time is the number of the motion of the celestial bodies'.[28] Astrological timing delineates a kind of a sacred space in motion. Images and sigils

---

[24] John Heydon, *Theomagia* (London, 1664), p. 178. This passage is in fact 'quoted' verbatim from Agrippa (*Three Books*, p. 371). The planets were assigned 'giving always the first hour of the day to the Lord of that day, then to every one by order'. This 'order' of the ruling Lords is the so-called Chaldean order of ascending speeds of the planets, Saturn deemed slowest and the Moon fastest.

[25] Kassell, *Medicine and Magic*, p. 48; citing MS Ashm 244, fols. 96–96v. The eighth heaven is the first 'sphere' in Cabalistic philosophy beyond the planetary spheres.

[26] MS Ashm 244, fols. 48, 91–92. For Ashmole's reading of these notes, see Lauren Kassell, 'Economy of Magic in Early Modern England', in *The Practice of Reform in Health, Medicine, and Science, 1500–2000: Essays for Charles Webster*, ed. Margaret Pelling and Scott Mandelbrote (Aldershot: Ashgate, 2005), pp. 43–57.

[27] MS Ashm 421, fol. 109v.

[28] Agrippa, *Three Books*, p. 238 no. 2; citing Proclus, *On Motion* 2 in Thomas Taylor, *Ocellus Lucanus* (Los Angeles, CA: Philosophical Research Society, 1976 edition), p. 86.

were the fruits of these delineated matrixes; gathered and incubated until ready to emerge forth to stand alone from the context that birthed them.

Sigils and images, I reaffirm, demonstrate that 'magic here clearly fused with astrology', but I would like to emphasise that this exemplifies how magic and astrology were interdependent and thus present interdisciplinary complexes of theories and activities.[29] Sigils represent a clearer instance upon this spectrum of interdependence, not simply one side of a fundamental and irreconcilable ideological split between 'magical' and 'scientific' forms of astrology. It is clearly in the light of both understanding underlying astral influences and storing and deploying such virtues in sigils and images – of knowledge facilitating both affective and efficacious action – that Ashmole praises the practice, for by 'Elections we may Governe, Order and Produce things as we please: Faber quisq; Fortunæ propriæ [sic]'.[30]

Early modern English almanacs commonly gave astrological advice on the propitious times to 'plough, sow, geld animals or fell timber'.[31] Similarly, astrological meteorology was so expected, it was remarked that an almanac without weather predictions was considered to be 'like a Pudding without Sewet, or a Christmas-pye without Plums'.[32]

Agricultural astrological images could be used to increase productivity, such as an 'Image of the Moon for the increase of the fruits of the earth, and against poysons, and infirmities of children, at the hour of the Moon, it ascending in the first face of Cancer'.[33] The Moon was considered to be particularly powerful in early modern matters of agriculture – even farmers untrained in the formal calculation and literate interpretation of the more mathematical learned forms of astrology observed moon-phases to time

---

[29] Capp, *Almanacs*, p. 21.

[30] Ashmole, *Theatrum Chemicum Britannicum*, p. 451. The Latin phrase Ashmole uses – which is clearly meant to be 'every man is the architect of his own fortune' – is attributed to the Roman patrician Appius Claudius Caecus (340–273 BCE).

[31] Capp, *Almanacs,* p. 63.

[32] Adam Martindale, *Country Almanack For the Year 1676* (London, 1676), sig. B 2.

[33] '… the figure of which was a woman cornuted, riding on a Bull, or a Dragon… or a Crab; and she hath in her right hand a dart, in her left a looking glass, clothed in white or green, and having on her head two Serpents with horns twined together, and to each arm a Serpent twined about, and to each foot one in like manner.' Agrippa, *Three Books*, p. 389.

their endeavours.[34] It should therefore not surprise us that, given this lunar patronage, an image of the Moon should aid so.

Similarly, images could be used to specifically manage resources more accurately and, arguably, learnedly. An image of the first decan or 'face' of Taurus depicted 'a naked man, an Archer, Harvester or Husbandman, and goeth forth to sow, plough, build, people, and divide the earth, according to the rules of Geometry'.[35] This function is also symbolically consistent, with Taurus being a sign of the Earthy triplicity.[36]

The use of astrological images against pests, such as those made by Ashmole (and recommended by Forman), is also found in an entomological apotropaic image from a (pseudo-)Paracelsian treatise on medical sigils, 'Englished' by occultist and medical researcher Robert Turner (1619–1664) in 1655.[37] Around an illustration of some kind of an arthropod, the text instructs:

> If such an Animal as follows be made of pure Iron, when Mars is Lord of the yeer, and the Sun enters the first degree of Scorpio; afterwards when Mars is in his own House in Aries... Then let it be applied in the hour of Mars : the House wherein it is hanged, it defendeth safe from all Scorpions; and all Serpents that are alive will flie out of it : it is a most excellent remedy against all venemous bitings... Let there be affixed a Ring of pure Gold to the Tayle thereof, that it may be worn hanging about the Neck with Head downwards. It is a certain Remedy to drive away all Flies from the bed where it is hanged.[38]

Timing and materials for its construction include both Signs and the metal ruled by Mars. Instructions for hanging are given in planetary hours and with specific directions concerning materials used to hang it, and even which way up and exactly where it should be hung.

Astrological images are most commonly understood as involving illustrations or engravings, such as those of the Moon and Taurus above, but representational magical images or effigies – formed in the image of its

---

[34] Astrologer and mathematician Henry Coley (1633–1704) even admitted that, arguably, all agricultural astrology really needed was the zodiac sign that the sun occupied and the phases of the moon. Coley, *Clavis*, sig. C7v–8.

[35] Agrippa, *Three Books*, p. 377.

[36] Lilly, *Christian Astrology*, p. 95; Coley, *Clavis*, pp. 10, 20–21.

[37] Forman listed (*inter alia*) magical objects used in medical treatments, to gain wealth, and to ward against vermin. MS Sloane 3822, fols. 6–19, 80–80v, 94–102.

[38] *Paracelsus of the Supreme Mysteries of Nature*, trans. Robert Turner (London, 1655), p. 148.

intended target of effect – also appear to be not uncommon. Ashmole cast lead miniatures of pests against infestations at a Saturn-Mars conjunction as mentioned above, and also even made effigies of genitalia as sigils of Mars and Venus 'against the pox &c'.[39] His notes on his pest-effigies read:

> The Figure of a Ratt was cast in Lead, and made in full proportion, but had noe Characters upon it. The Figure of the Mole was cutt in an Ovall Figure upon a Punchion of Iron, lying long-waies; over the back were these Characters set [astrological glyph for conjunction] [Leo]. under the Belly these [Seals of the Spirits of Mars and Saturn] And they were stamped upon Lead. The Figure of the Caterpillars & Flyes Fleas & Toades, were all made in full proportion, in little, & cast off in Lead, without Characters.[40]

Far from being the kind of glyphed medallions generally associated with early modern sigils, some of these sigils were miniature effigies of the pests they warded against. Ashmole's lead pests were not anomalous in early modern English magic. Written and spoken rat charms to oust rodents from one's home have a longer and wider European history.[41] These effigies have clear agricultural application, but such images were also utilised as a magical solution for this very real problem of wild nature encroaching on civilised domesticity.

From introducing astrological images used to affect the natural world, I would like to shift to considering the natural materials used to construct images. Let us firstly consider astrological metals. Understanding occult significances of metals can begin with their underground origins, 'wherein great treasures, and mighty store of wealth and Riches are hid', guarded by 'many terrene earthly Spirits' who 'do take care and custody thereof'; understandably then, those who 'digge Metals have the best knowledge of these Spirits'.[42] From their very beginnings, metals were connected with occult forces and agents, with chthonic mysteries and potencies.

Ashmole's pest castings in lead re-affirm the importance of using planetary metals – lead being considered the material of Saturn. These

---

[39] MS Ashm 431, fol. 136v. For more on these *penes et vulvae* cast', see MS Ashm 431, fol. 125v.

[40] MS Ashm 431, fol. 137.

[41] Wilson, *Magical Universe*, p. 20. See also Paul Cowdell, '"If Not, Shall Employ 'Rough on Rats'": Identifying the Common Elements of Rat Charms', in Jonathan Roper, *Charms, Charmers and Charming* (Basingstoke: Palgrave MacMillan, 2009), pp. 17–26.

[42] *Paracelsus of the Supreme Mysteries*, p. 52.

systematic correspondences – Saturn, lead; Jupiter, tin; Sun, gold; Venus, copper; Mercury, quicksilver (or, less commonly, alloys); Moon, silver – were occult standards, and were utilised to draw, contain and radiate the proper planetary virtues (and thus influence and affect) most effectively.[43]

Ashmole's records demonstrate these standards actually being observed in practical sigil-casting.[44] Even works dedicated to the medical application of such sigils contained detailed instructions and notes for practical metallurgy: *Supreme Mysteries*, for instance, noting the time and temperature it took to smelt iron for a sigil against gout, advised using 'a very strong fire, that the filings of Iron may be melted. For they will hardly melt, wherefore some Boras is to be added to them'.[45] The metallurgical know-how in such instructions also demonstrated a practical comprehension informed by contemporary occult philosophy of matter, for 'the Spirit of Mars is endued with a greater hardness then the other Metals; so that it doth not so easily melt and dissolve in the fire… in hardness and dryness it exceedeth all other Metals, both superior and inferiour : for it doth… resist the hammer…'[46] Far from being tangents on the nature of matter in a medical treatise, these fundamental qualities are specifically apprehended in the context of their effect on the human body.[47] The sharp, hard, dry, 'grievous' and harshly cleansing nature of Mars is understood in

---

[43] Agrippa, *Three Books*, pp. 75, 80, 83, 86, 89, 94, 258. These standards were shared by strictly astrological handbooks. Lilly, *Christian Astrology*, pp. 60, 64, 68, 72, 75, 79, 82. Likewise, the Paracelsian medical sigils of *Paracelsus of the Supreme Mysteries* also used these correspondences. *Paracelsus of the Supreme Mysteries*, p. 3. Hiebner's scientific reformist sigils swapped planetary governances of copper and tin between Venus and Jupiter, but were otherwise identical. Hiebner, *Sigillorum*, pp. 161, 165, 172.

[44] An election chart set for 11.30am on 27 July 1651, for instance, confirms Ashmole cast solar sigils in gold. Similarly, his anti-vermin sigils of Saturn and Mars used lead (the exemplary Saturnine metal), and his sigils of Jupiter and Mercury were made of Jupiterian tin. MS Ashm 374, fol. 117v (Sun); Ashm 431, fol. 137 (Saturn); Ashm 431, fol. 142v (Jupiter).

[45] *Paracelsus of the Supreme Mysteries*, p. 119. Emphasis added.

[46] *Paracelsus of the Supreme Mysteries*, p. 9.

[47] 'Whereas therefore it thus worketh in Metals, it sheweth that *it hath the same effect* in the bodies of men, that is, it produceth reluctancy; especially where it is taken for a disease not convenient, it greviously afflicteth the members with pain. Nevertheless, when it is taken applyed for wounds, such as do not exceed it own degree, it cleanseth and mundifieth them, &c.' *Paracelsus of the Supreme Mysteries*, p. 9. Emphasis added.

both the qualities of its metal and the effects such a metal produces biologically and medically. The powers of environmental influences, such as the virtues of metals, are brought to bear once more in a particularly human context.

Images could also be made by engraving onto suitable stones, which could be geologically specific – 'to Jupiter belong the Smaragdine [emerald], Sapphire, and Amethist' – or more qualitatively descriptive, as 'all black, wild, and dark Stones belong to Saturn'.[48] Planetary, zodiacal and decanate virtues might be combined in a single image engraved onto a stone, such as one made 'in the hour of Mars, Mars being in the second face of Aries, in a Martial stone, especially in a Diamond'.[49] Once more, a material renowned for its hardness is attributed to Mars. Lapidary or 'mineral' correspondences were also common features in astrological handbooks, and such information could clearly be used to construct images as well as other magical objects.[50]

The use of astrological metals and stones obviously highlights the extent and significance of natural materials in astrological image magic. But there were other, arguably even more 'natural', substances utilised. Wax, an impressionable substance after all made by animals, was a common material for making magical images. Wax might be used like metal as a base to be impressed into, or even as a container for other materia: an image of the first lunar mansion was to be made upon an iron ring and then sealed in black wax.[51]

Given that marking or impressing pictures, symbols, ideograms and glyphs into suitable materials at suitable times was thought so central, it should not surprise us that there exist instructions for directly marking

---

[48] Hiebner, *Sigillorum*, p. 167, 164.

[49] 'The form of which was a man armed, riding upon a Lyon, having in his right hand a naked sword erected, carrying in his left hand the head of a man; they report, that an Image of this kind rendreth a man powerfull in good and evill, so that he shall be feared of all; and whosoever carryeth it they give him the power of enchantment, so that he shall terrifie men by his looks when he is angry, and stupifie them…' Agrippa, *Three Books*, p. 385.

[50] Indeed, the 'Minerals' field often covered planetary governances of metals and stones. See Lilly, *Christian Astrology*, pp. 60, 64, 68, 72, 75, 79, 82.

[51] '…for the destruction of some one, they made in an Iron ring, the Image of a black man in a garment made of haire, and girdled round, casting a small lance with his right hand; they sealed this in black wax, and perfumed it with liquid Storax, and wished some evil to come' Agrippa, *Three Books*, p. 392.

natural materials. Agrippa describes an image of the twenty-fourth lunar mansion,

> ... for the multiplying of herds of cattle, they took the horn of a ram, bull or goat, or of that sort of cattle which they would increase, and sealed in it burning with an iron seal, the image of a woman giving suck to her son, and they hanged it on the neck of that cattle who was the leader of the flock...[52]

Similarly, 'In the five and twentieth [mansion], for the preservation of Trees and Harvests, they sealed in the wood of a Fig-tree, the image of a man planting, and they perfumed it with the flowers of the Fig-tree, and did hang it on the tree'.[53]

As these instruction suggest, once constructed, one of the further crucial ways in which an image was consecrated or magically activated was through fumigation using natural materials which contained and thus conferred particular occult virtues. Incense was a crucial component in conjuring spirits, themselves denizens of a vast magical ecology.[54] Ashmole's notes for 5 May 1678 reveal he cast two batches of sigils at 8.15 and 8.30 pm, before the five total sigils were 'fumed at 8.35' P.M.', giving us a better idea of both his general process and the length of time casting might take.[55]

One could use three main means to magically fumigate your image: by material type, by plant identity, and by 'savour'. Material type was the broadest rule-of-thumb for planetary incenses: working by the part of the plant being used. This was found in workbooks of magicians such as the texts attributed to Arthur Gauntlet as well as more formal works of occult philosophy.[56]

---

[52] Agrippa, *Three Books*, p. 393.

[53] Agrippa, *Three Books*, p. 393.

[54] '... let no man wonder how great things suffumigations can do in the Air, especially when he shall with Porphyrius consider, that by certain vapours exhaling from proper suffumigations, airy spirits are presently raised, as also thunderings, and lightnings, and such like things.' Agrippa, *Three Books*, p. 129.

[55] MS Ashm 421, fol. 107v.

[56] *Gauntlet*, p. 257 n. 334–340. '... to Saturn are appropriated for fumes all odoriferous *roots*, as Pepper-wort root, &c. and the Frankincense tree: to Jupiter odoriferous *fruits*, as Nutmegs, Cloves: to Mars all odoriferous *wood*, as Sanders [sandalwood], Cypress, Lignum-balsaim [lignum balsam], and Lignum-aloes: to the Sun, all *Gums*, Frankincense, Mastick, Benjamin, Storax, Laudanum [labdanum, i.e. Cistus], Amber-gryse [ambergris], and Musk; to Venus *Flowers*, as

One could use strict astrological governances of plants, and most contemporary herbals offered such classifications. To continue our Martial examples: 'The Hearbs we attribute to [Mars] are such as come neare to a rednesse, whose leaves are pointed and sharp, whose taste is costick and burning, love to grow on dry places, are corosive and penetrating the Flesh and Bones with a most subtill heat...'[57] Such governances were themselves determined by morphology, scent, behaviour and sympathies.

Astrological signature and elemental-humoural properties were also covered by 'savours'. Thus the traditional grimoire of *Sepher Raziel* reports: 'Hermes said of suffumigations, "Aries, Leo and Sagittarius hold each choleric spicy & bitter; Taurus, Virgo and Capricorn melancholic and are styptic [astringent]; Gemini, Libra & Aquarius sanguine and are sweet; Cancer, Scorpio & Pisces phlegmatic and are of a salty savor"'.[58] Handbooks such as Lilly's *Christian Astrology* give these same 'savours' for the planets and signs.[59]

Literal sensory data was assigned and combined with (what modern thinkers would probably consider 'figurative') descriptors of affectivity. As Agrippa described, and was frequently quoted by contemporary magical practitioners, 'in every good matter, as love, goodwill, and the like, there must be a good fume, odoriferous, and precious; and in every evil matter, as hatred, anger, misery, and the like, there must be a stinking fume, that is of no worth.'[60] Indeed, how pleasant or foul a thing smelled was considered a testament to its nature. There was a signature, a notion that the 'perceived-outside' somehow reflected the 'essential-inside' to scent as well as morphology.[61]

As we saw from the lunar mansion images sealed in animal horn and bark, such images might be placed in significant positions; in those cases, around the neck of the head of the herd or on the tree. These are instances

---

Roses, Violets, Saffron, and such like: to Mercury all *Pils* [peels] of Wood and fruit, as Cinnamon, Lignum Cassia, Mace, Citron pill [lemon peel], and Bayberries, and whatsoever *seeds* are odoriferous; to the Moon the *leaves* of all Vegetables, as the leaf Indum, the leaves of the Myrtle, and Bay-tree.' Agrippa, *Three Books*, p. 132.

[57] Lilly, *Christian Astrology*, p. 75.

[58] MS Sloane 3826 fol. 33v; translated in Karr and Skinner, *Sepher Raziel*, p. 193.

[59] Lilly, *Christian Astrology*, pp. 59, 63, 67, 71, 75, 79, 82.

[60] Agrippa, *Three Books*, p. 132.

[61] For more on the doctrine of signatures, see Jakob Böhme, *Signatura rerum, or, The signature of all things shewing the sign and signification of the severall forms and shapes in the creation* (London, 1651).

of the actual deployment of images, and the enterprising image magician could deploy them 'diversly according to the vertues thereof'.[62] However, Agrippa also offers us a fascinating insight into the more general manner in which images could be further empowered by such locative deployment:

> ... they that endeavour to procure love, are wont to bury for a certain time the instruments of their art, whether they be rings, images, looking glasses, or any other, to hide them in a stewhouse [brothel], because in that place they will contract some venereal [i.e. Venusian] faculty, no otherwise than things that stand in stinking places, become stinking, and those in an aromatical place, become aromatical, and of a sweet savour.[63]

By occult doctrines of contagion and exposure utilised in receiving, storing and manipulating astral virtue, these locations were themselves places of power which could be used to charge magical objects. The correspondence lists of locations found in every astrological handbook were not merely for decoding information from charts – although this was incredibly useful, and facilitated part of contemporary cunning-folks' stock trade in finding lost or stolen property[64] – it was also for apprehending and utilising loci and materials to manipulate, store, and transmit astral virtues.

Having considered images for environmental, ecological, or geographical utility, and indeed the natural sources of materia magica, this brief survey will finish by examining magical images and iconography of nature.

> They made another Image of Venus, the first face of Taurus or Libra or Pisces ascending with Venus, the figure of which was a little maide with her hair spread abroad, cloathed in long and white garments, holding a Laurell[,] Apple, or flowers in her right hand, in her left a Combe. Its reported to make men pleasant, jocand, strong, chearfull and to give beauty.[65]

---

[62] 'Sometimes they hang them or binde them to the body; Sometimes they bury them under the Earth, or a River; sometimes they hang them in a Chimny over the smoak, or upon a tree that they be moved by the wind; sometime with the head upward, & sometimes downward; sometimes they put them into hot water, or into the fire.' Agrippa, *Three Books*, p. 400.

[63] Agrippa, *Three Books*, p. 144. Emphasis added.

[64] Davies, *Popular Magic*, p. 100.

[65] Agrippa, *Three Books*, p. 387.

Let us examine the occult comprehensions, analyses and utilities of the two pictured natural materials – the laurel and the apple.

Nicholas Culpeper (1616–1654) reports the bay tree as under Jupiter, lending the amulet its jovial character.[66] Culpeper also points out that the bay is sometimes attributed to the Sun 'and resisteth Witchcraft very potently, as also al the evil old Saturn can do to the Body of Man'.[67] Such is a clear instance of cure by planetary contrary, as well as the consideration of witchcraft and supernatural disease alongside the natural magic of the planets. Saturn was also the planet that most powerfully ruled the cold and dry sorrowful black bile; that humour responsible for melancholic symptoms that modern thinkers might call 'depression'.[68]

Agrippa gives two examples of bay leaves being used to wrap up magical objects, usually to preserve solar virtues – again, frequently used against Saturnine melancholy. Agrippa also tells us bay has Martial virtues, and although he cautions it is 'poisonous, by reason of too much heat', considered in light of equilibrating health from the unbalanced cold humours of melancholic dyscrasia, this heat could have practical benefits.[69] Finally, Hiebner attributed the laurel to Saturn for the very reason that it commanded against it so well, exemplifying cure by command through planetary rulership to banish its humours and passions.[70] In short, bay laurel empowered magical objects, and defended against Saturnine disease, misfortune, bewitching and melancholy. It seems an image involving bay laurel did the same.

With similar anti-depressant effects, Culpeper's appraised 'the sweet Apples as the Pippin and Pearmain, help to dissolve Melancholly humors, and to procure Mirth'.[71] Likewise, Robert Turner gives this same appraisal in his *Botanologia*, and Henry Butts, in his treatise on diet, gives their first use as to 'Comfort the hart'.[72] The heart was of central importance in early modern comprehensions of joy and sorrow, considered the very shaper of

---

[66] Culpeper, *English Physician*, p. 11.

[67] Culpeper, *English Physician*, p. 12.

[68] For more on the polyvalency, problematics, and historiography of melancholy, see Angus Gowland, 'The Problem of Early Modern Melancholy', *Past and Present* 191 (May, 2006), especially pp. 78–79.

[69] Agrippa, *Three Books*, p. 89.

[70] Hiebner, *Sigillorum*, p. 23.

[71] Culpeper, *English Physician*, p. 5. For more on the magical properties of apples, see Chapter 6.

[72] Henry Butts, *Dyets Dry Dinner* (London, 1599), f. 19.

the passions.[73] Crucially therefore, in the combination of laurel and apple,
we see occult principles of both commanding or nullifying ill effects *and*
bolstering and encouraging positive ones; the malefic is banished and the
benefic beckoned and strengthened.

The apple has been called the most magical of the fruit trees: sorcerers
could work curses, charms and divinations from it and its fruit.[74] Apples
are, of course, Venusian.[75] A Scriptural symbol of temptation, the most
popular magical use of apples was for love – the workbook-cum-grimoire
attributed to Arthur Gauntlet, for instance, contains eighteen distinct love
spells involving apples – and involved mostly writing a charm or drawing
characters on them, and then getting the target of one's affections to eat
it.[76] A Scottish account also tells us that in 1601 one David Roy attempted
to seduce his employer's daughter 'by the dobbing on ane apple and
infusing of ane portioun of his awn nature [i.e. semen] in it', an interesting
example of using natural materials pertaining to love to secure it.[77]

Natural plant material thus becomes a medium and delivery system.
The two key components of early modern love magic were to heat and to
sweeten: love being considered literally as well as figuratively sweet.
Agrippa also considers apples as one of the 'things under Jupiter' and
'amongst Elements, are the Aire: amongst humors, blood, and the spirit of

---

[73] See Fay Bound Alberti, *Matters of the Heart: History, Medicine, and Emotion*
(Oxford: Oxford University Press, 2010).
[74] K. M. Briggs, 'Some Seventeenth-century Books of Magic', *Folklore* 64, no. 4
(1953): p. 456.
[75] Culpeper, *English Physician*, p. 5.
[76] Rankine, *Gauntlet*, pp. 305–8.
[77] Julian Goodare, 'Men and the Witch-Hunt in Scotland' in *Witchcraft and
Masculinities in Early Modern Europe* ed. Alison Rowlands (Basingstoke:
Palgrave MacMillan, 2009), p. 159. It should also be noted that in early modern
thought 'semen, as Galen had explained, was concocted out of blood'. Nogah
Arikha, *Passions and Tempers: A History of the Humours* (New York:
HarperCollins, 2007), pp. 164–65. It thus had very particular humoural and
passional connotations, most notably 'that those that are of a sanguine
Complexion, are generally very Amorous'. James Ferrand, *Erotomania or A
Treatise Discoursing of the Essence, Causes, Symptomes, Prognosticks, and Cure
of Love or Erotique Melancholy* (Oxford, 1640), p. 141 For more on seed and
blood, particularly as related to love and love sickness, see Lesel Dawson,
Lovesickness and Gender in Early Modern English Literature (Oxford: Oxford
University Press, 2008), pp. 25–26, 85–86, 165–66, 209.

life' – which also spoke to both the inherent joviality and lasciviousness accredited to the sanguine Airy humoural complexion.[78]

Significantly then, the sweetness of the apples not only countered sour black bile in actuality, but could summon such a counteracting physical effect by its mere image, by the impression of its correspondences of virtue upon the imagination.[79] Significantly then, images of plants and natural materials with appropriate virtues were used as effective parts of an overall magical operation in astrological image magic. Giambattista della Porta (1535–1615) cites that the Roman naturalist Pliny the Elder claimed 'that an herb which grows in the head of an Image, being wrapt in a cloth, is good for the Head-ach'.[80] Here we find a similitude of origin – both herb and ache emerge in the head, thus the latter can therefore be affected by the former. Crucially, the virtual or representative head in an image, possessing a similitude of a head, might be as magically valid as an actual human head. Images affect the virtues of organic natural materials.

In contrast to the efforts to wall off human spaces like beds from unwanted animal visitations, agricultural utilities of astrology actually further affirm the advantages of considering humanity as itself an expression of the natural environment, in both harmony and conflict with annual cycles of breeding seasons, harvests and changing weather. Astrology of this ilk sought to understand, predict, and avoid natural disasters such as flood, famine and plague. Astrologers not only attempted to explain the forces influencing relationships between humans and their landscape, crops and animals, but explore and refine the best techniques and methodologies to harness such forces towards desired advantageous outcomes.

'Natural' astrology therefore didn't necessarily mean 'scientific', in that it too could have magical applications. Yet these occult dimensions also didn't necessarily make it 'unscientific', as Hiebner's sigil precision reforms demonstrate. It is not only the *astrologia sana* of Francis Bacon (1561–1626), supported by both amateurs such as Samuel Jeake (1623–1690) and professional astrologers like John Gadbury (1627–1704), which 'saw astrology as comparable to other forms of natural causation'.[81]

---

[78] Agrippa, *Three Books*, p. 86. See also previous note about blood, semen and apples.

[79] It seems significant one of Agrippa's main examples of the material somatic effects of the imagination was gustative: 'some when they hear any one name soure things, their tongues waxeth tart'. Agrippa, *Three Books*, p. 201.

[80] Porta, *Natural Magick*, p. 17.

[81] Hunter and Gregory, *Jeake*, p. 14.

Indeed, in considering the rationality of attempting to operationalise a knowledge of natural causation to one's advantage, such as with astrological images, we come to a fittingly agricultural analogy. Sigils and images represented the ripe fruit of astral power preserved in matter. Occult philosophy and magical practice framed images 'in which appropriate heavenly influences were caught like fruit as they fell and stored up for use when needed'.[82] As Porta commented, 'as in Husbandry, it is Nature that bring forth corn and herbs, but it is Art that prepares and makes way for them'.[83] Indeed, 'a canny operator could catch the fruits of the stars just as they would pick the blooms of medicinal plants at the proper time of their blossoming'.[84]

Astrological image magic highlights how theoretically and practically reliant magic and astrology were on each other. I conceive of this in terms of mutually necessary principles of operation and organisation of the cosmos. Operational principles such as sympathy and organisational principles of specific astrological categories were necessary for both astrology and magic. At the very least, the categories used for divination and other epistemological endeavours – especially geographical correspondences such as the planetary locations – could be used just as effectively to perform operative enchantment and spellcraft. Not only did the celestial point to the worldly, but the worldly pointed back.

---

[82] Thomas, *Religion and the Decline of Magic*, p. 759.
[83] Porta, *Natural Magick*, p. 2.
[84] Cummins, *The Starry Rubric*, p 54.

NOTES ON CONTRIBUTORS

Gerardina Antelmi is an independent scholar, whose academic interests lie in Medieval Literature, Linguistics and Women's Studies. She is also a passionate creative writer. On graduating from Genoa University in Modern Languages and Literature, she began teaching, and her chosen career has taken her far from her native Italy. In 2003 she was appointed to Guangdong University of Foreign Languages, Guangzhou, China, where she taught Italian Literature and Language. Gerardina subsequently took up post at Cardiff University, where she completed her PhD on medieval oneiric literature, with a focus upon the dream framework in the works of Chaucer. Eager to experience other cultures, she moved to Poland where she lectured at Adam Mickiewicz University in Poznań, and then on to Split University, Croatia, finishing there in 2015. She has regularly contributed to the Italian feminist review *Marea*, and is an experienced translator of longstanding; her publications include *Géographie sacrée du monde grec. Croyances astrales des ancienc grecs* by Jean Richer. Three of her short stories have been awarded prizes in Italy. Her present research project concerns medieval terminology related to dreams and states of consciousness in medieval literature.

Juan Antonio Belmonte (Murcia, 1962) is a Research Professor of Astronomy at the Instituto de Astrofísica de Canarias (Tenerife, Spain) where he investigates exoplanets and cultural astronomy. He received in 2012 the *Carlos Jaschek Award* for his contributions to this discipline. He was the Director of the Science and Cosmos Museum of Tenerife from 1995 to 2000, President of the European Society for Astronomy in Culture (SEAC) from 2005 to 2011 and President of the Spanish Time Allocation Committee of the Canarian observatories for nearly a decade. He is now Advisory Editor of the *Journal for the History of Astronomy* and a member of the board of C41 History of Astronomy of the IAU. In recent years he has undertaken extensive research on the astronomical traditions of ancient civilizations, concentrating in the ancient Mediterranean, notably in Egypt. He is the editor of *In search of cosmic order: selected essays on Egyptian archaeoastronomy*, published by the Supreme Council of Antiquities Press in 2009, and the author of *Pirámides, templos and estrellas: astronomía and arqueología en el Egipto Antiguo*, published by Crítica in 2012. He has been the Spanish representative in the Egyptian-Spanish Mission for the Archaeoastronomy of Ancient Egypt for nearly a decade and is now a member of the Spanish Archaeological Mission at Herakleopolis Magna, investigating the local sacred landscape. jba@iac.es

Alexander Cummins obtained his doctorate from the University of Bristol, where he studied the history of early modern English magical approaches to the emotions. He has presented and published on topics such as the material history amulets, comparative millenarianism, botany and herbalism in the grimoires, humoural theory and necromancy. His first history book, *The Starry Rubric:*

Seventeenth-century English Astrology and Magic, published by Hadean Press in 2012, examines the environmental, political and social functions of astrology and magic in this period.

Edina Eszenyi is Resident Art Historian of the *Rome Art Program*, a New York-based non-profit international HE institution providing on-site training for artists and art students in Italy (www.romeartprogram.org). She holds a PhD in Medieval and Early Modern Studies from the University of Kent Canterbury, UK, two MA diplomas (Art History and English Studies) from the Pázmány Péter Catholic University of Hungary, and a postgraduate research MA in Medieval Studies from Central European University. She lives and works in Rome. She specializes in the historical and artistic aspects of angelology, with a particular interest in the Fall of the Angels tradition.

Shon D. Hopkin is an Assistant Professor of Religious Education at Brigham Young University. He earned his BS and MA from BYU in Ancient Near Eastern Studies with an emphasis on Biblical Hebrew, and his PhD in Hebrew Studies from the University of Texas at Austin. He researches and writes on the connection between ritual theory, biblical texts such as Leviticus and Psalms, and modern religious practices such as those exhibited by Mormons and others. His published works include: 'Representing the Divine Ascent: The Day of Atonement in Christian and Nephite Scripture and Practice', 'Salvation by Grace, Rewards of Degree by Works: The Soteriology of Doctrine and Covenants 76', 'Peter, Stones, and Seers', 'The Psalms as Signifiers of Sacred Time and Space', and *Mormonism: A Guide for the Perplexed* (with Robert L. Millet).

Harold H. Green (Hal), a graduate of Stanford University and Harvard Law School, is a retired lawyer who has been a student of ancient Maya cosmology, calendrics and culture since 1998. He is the author of three studies into the relationship of the zenith sun to Mesoamerican calendrics and sacred space orientation. His study entitled 'Cosmic Order at Chocolá: Implications of Solar Observations of the Eastern Horizon at Chocolá, Suchitepéquez, Guatemala' is published as Chapter 1 in *Archaeoastronomy and the Maya*, ed. Gerardo Aldana y Villalobos and Edwin L. Barnhart (Oxbow Books, 2014). He has attended the Texas Maya Meetings since 2000, is an Associate Member of the Society for American Archaeology, a Member of the European Association of Mayanists, and was a volunteer for the Proyecto Arqueológico Chocolá in Guatemala (2005–2006). He attended ISAAC's Oxford IX conference (Lima 2011) and was a presenter at The Solstice Project Seminar 'Chaco/Mesoamerica: Considerations of Parallel Expressions of Astronomy, Geometry and Latitude' (Commonweal 2013).

Scott Hendrix is an Associate Professor of History at Carroll University. He is the author of *The Impact of the English Colonization of Ireland in the Sixteenth Century* (Edwin Mellen, 2012), *Riot and Resistance in County Norfolk, 1646–1650*

(Edwin Mellen, 2012), and *How Albert the Great's Speculum Astronomiae Was Interpreted and Used by Four Centuries of Readers: A Study in Late Medieval Medicine, Astronomy, and Astrology* (Edwin Mellen, 2010). He is also the coeditor (with Christopher J. May) of *Integrative Mysticism* (Interdisciplinary Press, 2013) and (with Brian Feltham) *Rational Magic: Cultural Studies in Magical Traditions* (Interdisciplinary Press, 2010).

Stanisław Iwaniszewski is Professor of Archaeology at the Department of Postgraduate Studies in the National School of Anthropology and History (ENAH-INAH), Mexico City, and curator at the State Archaeological Museum in Warsaw. He specializes in the archaeology of identity, landscape archaeology, and archaeoastronomy. In 2015 he habilitated with a monothematic cycle of publications: *Theoretical and methodological aspects of research in archaeoastronomy.*

Kim Malville obtained his BSc in physics from California Institute of Technology and his PhD in radio astronomy and solar physics from the University of Colorado. He is presently Professor Emeritus in the Department of Astrophysical and Planetary Sciences at the University of Colorado. In 1997 he was a member of the team that revealed the world's oldest known megalithic astronomy at Nabta Playa near Abu Simbel in southern Egypt. In 2003 he was involved in the rediscovery of Llactapata, previously lost in a cloud forest near Machu Picchu. Books which he has written or edited include *A Feather for Daedalus; The Fermenting Universe; Prehistoric Astronomy in the Southwest; Canyon Spirits: Beauty and Power in the Ancestral Puebloan World; Ancient Cities, Sacred Skies: Cosmic Geometries and City Planning in Ancient India; Chimney Rock: the Ultimate Outlier; Pilgrimage: Sacred Landscapes and Self-Organized Complexity*, and *Machu Picchu's Sacred Sisters, Choquequirao and Llactapata: Astronomy, Symbolism, and Sacred Geography in the Inca Heartland.*

Joanna Popielska-Grzybowska holds a PhD in Egyptology (2007) from the Institute of Archaeology at the University of Warsaw. She is a philologist and archaeologist. She was a Lecturer at the University of Warsaw from 1998–2006, and has been a lecturer and adjunct from 01 November 2004 to the present at the Pułtusk Academy of Humanities, Poland. She has also been Head of Department of Ancient Cultures from 15 April 2011 to the present, and Head of Department of Cultures of Africa from 01 July 2009–14 February 2011.

Popielska-Grzybowska's publications include two forthcoming monographs, chapters in three monographs (2005, 2012, 2015), and 36 published articles, 18 forthcoming. She has been a member of the International Association of Egyptologists since 1998, a member of the Polish Society of African Studies since 2010, a charter member of the Society EDUCARE since 2013; *International Society for Ethnology and Folklore* since 2014; a charter member *Stowarzyszenie Uniwersytetów Dziecięcych* (=Society of Childrens' Universities) since 2015; and nominated correspondent researcher of the *Centro de História d'Aquém e d'Além-Mar /Portuguese Centre for Global History* since November 2015. In addition, she has organised ten international conferences; served as editor for 14 volumes of scientific publications, the whole series *Acta Archaeologica Pultuskiensia* included; is Founder of the University for Children (Pułtusk); and organises chatty lectures, workshops, and IMAX Cinema – educational programme.

BACK ISSUES OF CULTURE AND COSMOS

http://www.cultureandcosmos.org/backIssues.html

The Cross in Cygnus; **Angela Voss:** *The Astrology of Marsilio Ficino: Divination or Science?;* **Patrick Curry:** *Astrology on Trial, and its Historians: Reflections on the Historiography of 'Superstition'.*

Contents Vol. 5 no 1 (spring/summer 2001)
Demetra George: *Manuel I Komnenos and Michael Glykas: A Twelfth-Century Defence and Refutation of Astrology,* Part I; **Richard L. Poss:** *Stars and Spirituality in the Cosmology of Dante's* Commedia.

Contents Vol. 5 no 2 (autumn/winter 2001)
Arkadiusz Sołtysiak: *The Bull of Heaven in Mesopotamian Sources*; **Demetra George:** *Manuel I Komnenos and Michael Glykas: A Twelfth-Century Defence and Refutation of Astrology,* Part 2; **Garry Phillipson** and **Peter Case:** *The Hidden Lineage of Modern Management Science: Astrology, Alchemy and the Myers-Briggs Type Indicator.*

Contents Volume 6 Number 1 (spring/summer 2002)
Ari Belenkyi: *A Unique Feature of the Jewish Calendar - Dehiyot*; **Demetra George:** *Manuel I Komnenos and Michael Glykas: A Twelfth-Century Defence and Refutation of Astrology,* Part 3; **Germana Ernst:** *The Sky in a Room: Campanella's Apologeticus in defence of the pamphlet* De siderali fato vitando; **Tommaso Campanella:** *Apologia for the opuscule on* De siderali fato vitando.

Contents Volume 6 Number 2 (autumn/winter 2002)
Jesse Krai: *Rheticus' Poem* 'Concerning the Beer of Breslau and the Twelve Signs of the Zodiac'; **Anna Marie Roos:** *Israel Hiebner's Astrological Amulets and the English Sigil War*; **Nicholas Campion:** *Surrealist Cosmology: André Breton and Astrology.*

Contents Volume 7 Number 1 (spring/summer 2003) GALILEO'S ASTROLOGY
Nick Kollerstrom: *Foreword: Galileo as Believer*; **Nicholas Campion:** *Introduction: Galileo's Life and Work*; **Antonio Favaro:** *Galileo, Astrologer*; **Germana Ernst:** *Astrology and Prophecy in Campanella and Galileo*; **Nick Kollerstrom**; *Galileo as an Astrologer: Antonino Poppi: On Trial for Astral Fatalism: Galileo Faces the Inquisition;* **Guiseppe Righini:***Galileo's Horoscope for Cosimo II de Medici*; **Mario Biagioli:** *An Astrologico-Dynastic Encounter; Galileo's Correspondence; Galileo's Letter to Dini, May 1611; On the Character of Sagredo: Galileo's judgements upon his nativity; Galileo's Horoscopes for his Daughters; Rome, 1630*; **Bernadette Brady:** *Four Galilean Horoscopes: An Analysis of Galileo's Astrological Techniques; A Sonnet by Galileo.*

Contents Volume 7 Number 2 (autumn/winter 2003)
Günther Oestmann: *Tycho Brahe's Geniture*; **Bernard Eccles:** *Astrological physiognomy from Ptolemy to the present day*; **James Brockbank:** *Planetary signification from the second century until the present day*; **Julia Cleave:** *Ficino's Approach to Astrology as Reflected in Book VII of his Letters.*

Contents Volume 8 No 1/2 (spring/summer autumn/winter 2004)
Valerie Shrimplin *Organising INSAP*; **Rolf Sinclair** *Foreword: INSAP IV in Oxford: A Summary*; **Nicholas Campion** *Introduction: The Inspiration of Astronomical Phenomena*: **Hubert A. Allen, Jr.** *Hawkins' Way: Remembering Astronomer Gerald S. Hawkins*; **Hubert A. Allen, Jr. and Terry Edward Ballone** *Star Imagery in Petroglyph National*

Monument; **Mark Butterworth** *Astronomy and the Magic Lantern*; **Ann Laurence Caudano** *Sun, Moon, and Stars on Kievan Rus Jewellery (10th – 13th Centuries)*; **Nicholas Campion** *The Sun is God;* **Anne Chapman-Rietschi** *Cosmic Gardens*; **Deborah Garwood** *Paris Solstice*; **N. J. Girardot** *Celestial Worlds In the Work of Self-Taught Visionary Artists With Special Reference to Howard Finster's Vision of 1982*; **John G. Hatch** *Desire, Heavenly Bodies, and a Surrealist's Fascination with the Celestial Theatre*; **Holly Henry** *Bertrand Russell in Blue Spectacles: His Fascination with Astronomy*; Ronald Hicks *Astronomy and the Sacred Landscape in Irish Myth*; **Chris Impey** *Why Are We So Lonely?*; **Bernd Klähn** *The Aberration of Starlight and/in Postmodernist Fiction*; **Nick Kollerstrom** *How Galileo dedicated the moons of Jupiter to Cosimo II de Medici*; **Arnold Lebeuf** *Dating the five Suns of Aztec cosmology*; **Andrea D. Lobel** *Trailing the Paper Moon: Astronomical Interpretations of Exodus 12:1-2*; **Stephen C. McCluskey** *Wordsworth's 'Rydal Chapel' and the Astronomical Orientation of Churches*; **David Madacsi** *Sky: Atmospheres and Aesthetic Distance in Planetary and Lunar Environments*; **Daniel R. Matlaga** *A Journey of Celestial Lights: The Sky as Allegory in Melville's Moby Dick*; **Paul Murdin** *Representing the Moon*; **R. P. Olowin** *Robinson Jeffers: Poetic Responses to a Cosmological Revolution*; **David W. Pankenier** *A Brief History of Beiji (Northern Culmen)*; **Richard Poss** *Poetic Responses to the Size of the Universe: Astronomical Imagery and Cosmological Constraints*; **Barbara Rappenglück** *The material of the solid sky and its traces in cultures*; **Brad Ricca** *The Night of Falling Stars: Reading the 1833 Leonid Meteor Storm*; **Patricia Ricci** *Lux ex Tenebris: Etienne-Louis Boullée's Cenotaph for Sir Isaac Newton*; **Sarah Richards** *Die Planententheorie: its uses and meanings for the Saxon mining communities and the culture of the Dresden Court 1553-1719*; **William Saslaw and Paul Murdin** *The Double Apollos of Istrus*; **Petra G. Schmidl** *Dusk and Dawn in Medieval Islam; On the Importance of Twilight Phenomena with Some Examples of Their Representations in Texts and on Instruments*; **Valerie Shrimplin** *Borromini and the New Astronomy: the elliptical dome*; **Joshua Stein** *Cicero's Use of Astronomy as Proof of the Existence of the Gods*; **Antje Steinhoefel** *Art and Astronomy in the Service of Religion:Observations on the Work of John Russell (1745-1806)*; **Burkard Steinrücken** *An interpretation of the 'Sky Disc of Nebra' as an icon for a bronze age planetarium mechanism with parallels to the moving world-soul in Plato's* Timaeus; **Gary Wells** *Daumier and The Popular Image of Astronomy.*

Contents Vol. 9 no 1 (Spring/Summer 2005)

Gennadij Kurtik and Alexander Militarev *Once more on the origin of Semitic and Greek star names:an astronomic-etymological approach updated*; **Prudence Jones** *A Goddess Arrives: Nineteenth Century Sources of the New Age Triple Moon Goddess*; **Louise Curth** *Astrological Medicine and the Popular Press in Early Modern England.*

Contents Vol. 9 no 2 (Autumn/Winter 2005)

Marinus Anthony van der Sluijs *A Possible Babylonian Precursor to the Theory of ecpyrōsis*; **Liz Greene** *Did Orphic Beliefs Influence the Development of Hellenistic Astrology?*; **Ariel Cohen** *Astronomical Luni-Solar Cycles and the Chronology of the Masoretic Bible*; **Tayra Lanuza-Navarro** *An Astrological Disc from the Sixteenth Century*; **J.C. Holbrook** *Celestial Navigators and Navigation Stories.*

Contents Vol. 10 no 1 and 2 (Spring/Summer, Autumn/Winter 2006)

Lucia Dolce *Introduction: The worship of celestial bodies in Japan: politics, rituals and icons*; **Lucia Dolce** *The State of the Field: A basic bibliography on astrological cultic* *practices in Japan*; **Hayashi Makoto** *The Tokugawa Shoguns and Yin-yang knowledge (onmyōdō)*; **John Breen** *Inside Tokugawa religion: stars, planets and the calendar-as-*method; **Mark Teeuwen** *The imperial shrines of Ise:An ancient star cult?*; **Lilla Russell-Smith** *Stars and Planets in Chinese and Central Asian Buddhist Art from the Ninth to the Fifteenth Centuries*; **Matsumoto Ikuyo** *Two Mediaeval Manuscripts on the Worship of the Stars from the Fujii Eikan Collection*; **Tsuda Tetsuei** *The Images of Stars and Their Significance in Japanese Esoteric Buddhist Art*; **Meri Arichi** *Seven Stars of Heaven and Seven Shrines on Earth: The Big Dipper and the Hie Shrine in the Medieval* Period; **Gaynor Sekimori** *Star Rituals and Nikko Shugendô*; **Meri Arichi** *The front cover image: Myōken Bosatsu.*

Contents Vol. 11 no 1 and 2 (Spring/Summer, Autumn/Winter 2007)
Micah Ross *A Survey of Demotic Astrological* Texts; **Francis Schmidt** *Horoscope, Predestination and Merit in Ancient Judaism*; **Stephan Heilen** *Ancient Scholars on the Horoscope of Rome*; **Joanna Komorowska** *Philosophy among Astrologers* ; **Wolfgang Hübner** *The Tropical Points of the Zodiacal Year and the* Paranatellonta *in Manilius' Astronomica*; Aurelio Pérez Jiménez *Hephaestio and the Consecration of Statues*; **Robert Hand** *Signs as Houses (Places) in Ancient Astrology*; **Dorian Gieseler Greenbaum** *Calculating the Lots of Fortune and Daemon in Hellenistic Astrology*; **Susanne Denningmann** *The Ambiguous Terms* ἑῴα *and* ἑσπερία, ἀνατολή, *and* ἑῴα *and* ἑσπερία δύσις **Joseph Crane** *Ptolemy's Digression: Astrology's Aspects andMusical Intervals*; **Giuseppe Bezza** *The Development of an Astrological Term – from Greek* hairesis *to Arabic* ḥayyiz; **Deborah Houlding** *The Transmission of Ptolemy's Terms: An Historical Overview, Comparison and Interpretation.*

Contents Vol. 12 no 1 (Spring/Summer 2008)
Liz Greene *Is Astrology a Divinatory System?*; **James Maffie** *Watching the Heavens with a 'Rooted Heart': The Mystical Basis of Aztec Astronomy*; **J.C. Holbrook** *Astronomy and World Heritage.*

Contents Vol. 12 no 2 (Autumn/Winter 2008)
Mark Williams *Astrological Poetry in late medieval Wales: the case of Dafydd Nanmor's 'To God and the planet Saturn'*; **Scott Hendrix** *Choosing to be Human: Albert the Great on Self Awareness and Celestial Influence*; **Graham Douglas** *Luis Vilhena and the World of Astrology.*

Contents Vol. 13 no 1 (Spring/Summer 2009)
Josefina Rodríguez-Arribas *Astronomical and Astrological Terms in Ibn Ezra's Biblical Commentaries: A New Approach*; **Andrew Vladimirou** *Michael Psellos and Byzantine Astrology in the Eleventh Century*; **Marinus Anthony van der Sluijs** *The Dragon of the Eclipses—A Note*; **Patrick Curry** *Response to Liz Greene's 'Is Astrology a Divinatory System?'*

Contents Vol. 13 no 2 (Autumn/Winter 2009)
Liz Greene *Mystical Experiences Among Astrologers*; **Peter Pesic** *How the Sun Stood Still: Old English Interpretations of Joshua and the Leap Year*; **Doina Ionescu** *Virginia Woolf and Astronomy*; **Carlos Ziller Camenietzki and Luis Miguel Carolino** *Astrologers at War: Manuel Galhano Lourosa and the Political Restoration of Portugal, 1640–1668*; **Nick Campion** *Astrology's Role in New Age Culture: A Research Note*

Contents Vol. 14 no 1 and 2 (Spring/Summer, Autumn/Winter 2010)

Contents Vol. 15 no 1 (Spring/Summer 2011)

Contents Vol. 15 no 2 (Autumn/Winter 2011)

Contents Volume 16 No 1/2 (Spring/Summer Autumn/Winter 2012)

210

astronomy; **Aidan Foster**, *Hierophanies in the Vinland Sagas: Images of a New World*; **Inga Elmqvist Söderlund**, *Inspiration from antique heroic deeds: Hercules as an astronomer*; **Patricia Aakhus**, *Astral Magic and Adelard of Bath's Liber Prestigiorum; or Why Werewolves Change at the Full Moon*; **David Pankenier**, Astrology for an Empire: The 'Treatise on the Celestial Offices' (ca. 100 BCE); **Steven Renshaw**, *The Inspiration of Subaru as a Symbol of Values and Traditions in Japan*;b **Daniel Armstrong**, *Citing The Saucers: Astronomy, UFOs and a persistence of vision*; **Alberto Cappi**, *The concept of gravity before Newton*; **Paul Murdin**, *Artilleryman to head of state—how astronomy inspired Francois Arago*; **Paolo Molaro and Alberto Cappi**, *Edgar Allan Poe's cosmology in Eureka*; **Voula Saridakis**, *For 'the present and future happiness of my dear Pupils'": The Astronomical and Educational Legacy of Margaret Bryan*; **Michael Rowan-Robinson**, *The invisible universe*; THE ARTS: **Arnold Wolfendale**, *The Inter-Relation of the Visual Arts and Science in General and Astronomy in Particular*; **Lynda Harris**, *Changing Images of the Milky Way during the Greco-Roman and Medieval Periods*; **Lucia Ayala**, *The Universe in images: Iconography of the Plurality of Worlds*; **Tayra M. Carmen Lanuza-Navarro**, *Astrological culture before its public: the representation of astrology in Golden Age Spanish Theatre*; **Emily Urban**, *Depicting the Heavens: The Use of Astrology in the Frescoes of Rome*; **Michael Mendillo**, *The Artistic Portrayal of the Medicean Moons in Early Astronomical Charts, Books and Paintings*; **Rolf Sinclair**, *Howard Russell Butler: Painter Extraordinary of Solar Eclipses*; **Beatriz Garcia, Estela Reynoso, Silvina Pérez Alvarez and Rubén Gabellone**, *Inspiration of Astronomy in the movies: a history of a close encounter*; **Gary Wells**, *The Moon in the Landscape: Interpreting a Theme of 19th Century Art*; **Clive Davenhall**, *The Space Art of Scriven Bolton*; **Matthew Whitehouse**, *Astronomical Organ Music*; **Aaron Plasek**, *Between Scientists, Writers and Artists: Theorising and Critiquing Knowledge-Production at the Interstices between Disciplines*; ARTISTS: **Merja Markkula**, *The Way I See the Stars: fibre art inspired by astrobiology*; **Govinda Sah**, *Beyond the Notion*; **Gisela Weimann**, *Above all the stars*; **Courtney Wrenn**, *Nebulae (emission / absorption)*; **Toby MacLennan**, *Presentation of Playing the Stars*; **Felicity Spear**, *Extending vision: sky-situated knowledge and the artist's eye*; **Vanessa Stanley**, *Surveillance-Surveillance-Surveillance*; **Jim Cogswell**, *Molecular Delirium*.

Contents Vol. 17 no 1 (Spring/Summer 2013)

Contents Vol. 17 no 2 (Autumn/Winter 2013)

Fabio Silva: *Land, Sea and Skyscape: Two Case Studies of Man-made Structures in the Azores Islands*

Contents Vol. 18 no 1 (Spring/Summer 2014)
César Esteban, *Struggling for Interdisciplinarity: Reflections of an Astrophysicist Working in Cultural Astronomy*; **Ronald Hutton,** *Prehistoric British Astronomy: Whatever Happened to the Earth and Sun?*; **Nick Kollerstrom,** *Galileo and the Astrological Prophecy of Manuel Rosales*; **Clive Davenhall,** *Dr Katterfelto and the Prehistory of Astronomical Ballooning*; **Nicholas Campion,** *Celestial Art: An Interview with Geoff MacEwan.*

Contents Vol. 18 no 2 (Autumn/Winter 2014)
Roger Beck, *The Ancient Mithraeum as a Model Universe. Part 2*; **Helena Avelar and Charles Burnett,** *The Interpretation of a Horoscope Cast by Abraham the Jew in Béziers for a child born on 29 November 1135: An Essay in Understanding a Medieval Astrologer*; Lindsay Starkey, *Creation, Providence, and the Limits of Human Knowledge of the World: Mellin de Saint-Gelais and John Calvin on Astrology*; **Scott Hendrix,** *The Contextual Rationality of Galileo's Astrology*; **Richard Angelo Bergen, Paradise Lost and the Descent of Urania: from Astrology to Allegory** R. Hakan Kirkoğlu, Ilm-i nudjum *and 18th century Ottoman Court Politics*; **Graham Douglas,** *Trystes Cosmologiques: When Lévi-Strauss met the Astrologers.*

Contents Vol. 19 no 1 & 2 (Autumn/Winter and Spring/Summer 2015)
Mike Harding, *The Meanings of Magic*; **José Manuel Redondo,** *The Celestial Imagination: Proclus the Philosopher on Theurgy*; **Liz Greene,** *The God in the Stone: Gemstone Talismans in Western Magical Traditions*; **Claire Chandler,** *Investigating the Magical Practice found in PGM (Greek Magical Papyri) XIII*; **M. E. Warlick,** *Alchemy and the Transgendering of Mercury*; **Karen Parham,** *Teleological and Aesthetic Perfection in the* Aurora Consurgens; **Alison Greig,** *Angelomorphism and Magical Transformation in the Christian and Jewish Traditions*; **Christine Broadbent,** *Celestial Magic as the 'Love Path': The Spiritual Cosmology of Ibn 'Arabi*; **Hereward Tilton,** *Bells and Spells: Rosicrucianism and the Invocation of Planetary Spirits in Early Modern Germany*; **Joscelyn Godwin,** *Astral Ascent in the Occult Revival*; **Sue Lewis,** *The Transformational Techniques of Huber Astrology*; **Jane Burton,** *Ancient Necromantic Rituals in Contemporary Celestial Magic*; **Lilan Laishley,** *South Indian Ritual Dispels Negative Karma in the Birth Chart.*

Lightning Source UK Ltd.
Milton Keynes UK
UKOW07f1242120717
305181UK00011B/70/P